Bringing In The Hay

David A. Asson

A nostalgic history of agriculture's most romantic crop

Doubletree Advantage Corp.
Works In Nostalgic History
7380 SW 102nd Avenue Beaverton, Oregon 97008
Tel 503-913-7342
David@DoubletreeAdvantge.com

Library of Congress Control Number: 2003094179

ISBN: 0-9719982-0-5

Printed in Portland, Oregon by Dynagraphics, Inc.
Bindery by Lincoln & Allen, Portland, Oregon

Doubletree Advantage Corp.
Works In Nostalgic History
7380 SW 102nd Avenue   Tel   503-913-7342
Beaverton, Oregon 97008   Email David@DoubletreeAdvantage.com

Dedicated to my Father, Albert,
Who arguably was the best high boom hay stacker
On the Minidoka Project

With special appreciation to my wife, Carolyn
and my daughter Kathy Sanders for hours of editing
and to Ivar Nelson, Director of the University of Idaho Press
for invaluable comments and suggestions.

Table of Contents

		Page
PROLOGUE	CHANGING TIMES	7
CHAPTER 1	NOURISHMENT	11
CHAPTER 2	MAKING HAY	17
CHAPTER 3	SICKLES, SCYTHES AND MOWERS	23
CHAPTER 4	RAKES AND TEDDERS	31
CHAPTER 5	WAGONS AND LOADERS	39
CHAPTER 6	HAY MOWS AND TROLLEY CARRIERS	55
CHAPTER 7	COLOSSAL STACKERS	63
EPILOGUE	KEEPING TIME	87
REMEMBERING		89
APPENDICES		107
GLOSSARY		121
BIBLIOGRAPHY		123

I know of no pursuit in which more real and important services can be rendered to any country than by improving its agriculture, its breed of useful animals, and other branches of a husbandman's cares.

George Washington 1732 - 1799 U.S. General, President. Letter of July 20, 1794

Changing Times

Al-fasfasah touched my life at an early age. Many years passed, though, before I learned that this ancient and melodious Arabic word was the derivative name source of the cherished crop I knew as alfalfa. In original context, al-fasfasah meant 'best fodder'. Without question, alfalfa has been the favored animal feed throughout a grand history dating back to at least 490 BC. Industrious Persia of Southwest Asia is credited as being first to harvest the succulent nutrient. Along with munitions, the Persian militia transported its seed across the Mediterranean to Greece to cultivate an enduring sustenance ever vital to the horses and cattle of their invading armies.

But alfalfa – hay in the generic sense – is more than fodder. So much more. Plain, ordinary-

Martin Johnson Heade 1819 - 1904
Nineteenth century American painter

Normandy Hay Stacks by James Scoppettone
Gallery located in Santa Cruz, California

as-rain-at-the-wrong-time, hay is in my world the *romantic icon of all agricultural plantings.* Readers who lived the haying experience, think back. Friends who touched farming during weekend visits to Grandpa's farm, reflect.

Consider the stirring scenes famous artists painted over the years of delightful haying motifs. If not hay, what other produce captured favor so resoundingly? Clearly, Messrs. Scoppettone and Heade felt the rush of hay season in order to create their wonderful portrayals.

If these paintings fail to influence you to share my romantic contention, perhaps Roget's *'A Summer's Effort'* or Dristad's *'Wagons of Plenty'* will be more compelling.

My friend Trav with the Bitterroots in background

Or perhaps the purest aficionado of all – the cowboy. Slide imaginarily behind his saddle as he tends his hungry stock, tonguing a common stem between singing lips. Wavy patterns of windswept wheat may challenge the fluid dance of a neighboring field of timothy but there is no contest in the field of aroma. The sweet scent of freshly mown hay, wafting in summer air, has no equal. Even in its ultimate captured state, hay put up by the skilled hands of crafty stackers graces rural countrysides in monumental style. *What other harvest ever stirred such passion?*

To be sure, harvest atmosphere was sometimes stifling. Dust and chaff itched and irritated beneath sweaty shirts and loose trousers. But the satisfaction gained in walking a spotlessly mown and cleared field or watching the delicately balanced rise of a new stack salved spiritual ointment over each laborer's worn physique.

Too passionate? Maybe. I do admit inventing Roget and Dristad. *'A Summer's Effort'* and *'Wagons of Plenty'* are works of my imagination. But don't you think they should have been painted?

Sadly, much of the romance of old fashion harvesting has been lost to advancement.

Production-minded inventors developed labor-saving fuel-powered machines to speed up the harvest process. Scythes and barrel rakes were replaced with mowers and swathers. Balers and self-propelled harvesters displaced pitchforks, wagons and derricks. Modern efficiency necessarily feeds a faster world. So it is. Life transitions in the hay field as it does in all human endeavor.

Still the romance need not be lost entirely. Not unless we let it go. If we hold on, it will linger in photo, prose and painting, in song and poem and in the cheerful hearts of all who were, or would have liked to have been, the luckiest of laborers – the hay stackers.

Remembering

From rich, loam fields of dusty lore
Rise grand reflections of early ways;
When country passion nurtured crops
On horse drawn iron of gone by days.

Russets picked 'neath vine to basket;
Soft gold wheat bound by McCormick;
Pintos sieved in woven burlap;
Windrows raked in wait of derrick.

Now abandoned sun baked treasure
Stands forsaken to remind us
How chores performed in routine duty
Reaped the memory that ever binds us.

Pardon the dust, the sweat, the ragweed and endless rows of shocks. Sink into memory. Follow the scent. Mentally loop a scythe. Heft a three-tined spike one more time. Ride longingly atop the last load over the canal, past the old swimming hole to the feed yard. Soar untamed on a hanging derrick cable tire swing. Chew one more drying stem. Glimpse back

Near Middleton, Idaho

to that wonderful, simple world. Remember how it was – or how it may have been – that delightful, care free, childlike experience we christened *Bringing In The Hay.*

9

There seem to be but three ways for a nation to acquire wealth. The first is by war, as the Romans did, in plundering their conquered neighbours. This is robbery. The second by commerce, which is generally cheating. The third by agriculture, the only honest way, wherein man receives a real increase of the seed thrown into the ground, in a kind of continual miracle, wrought by the hand of God in his favor, as a reward for his innocent life and virtuous industry.

Benjamin Franklin (1706–90), U.S. statesman, writer. Positions to Be Examined Concerning National Wealth, written 4 April 1769.

CHAPTER 1

Nourishment

Types – Yields – Nutrients

Alfalfa, timothy, sweet clover, bluegrass, sudan, orchard grass, fescue, rye, and bermuda grass are listed among the roughages. Legume roughages, as they are technically known, come in both perennial and annual varieties making alfalfa a relative of peas, beans, peanuts and soybeans. This work deals only with hay types put up loose in outdoor stacks for later feeding to growing stock – particularly alfalfa, timothy, red clover and meadow grass. Meadow grass is a collective term for a mix of many local grasses such as brome, johnson, bluestem and indian grass.

Medicago Sativa, or alfalfa, as it is commonly known, ranks at the head of the list for hay crop productivity. It is often the ideal from which nutrient comparisons are made. A table taken from a book published in 1947 illustrates the statistics for these three varieties, grown widely in the Western States.

In this table, alfalfa excels in each category

yet clover and timothy rank well and are very popular as are the meadow grasses. Yield and nutrient content vary according to geography, soil type, weather and many other factors affecting agriculture. Understandably, any planting may eclipse another given the proper set of variables. This may account for the difference in many growers' opinions of worth and popularity. As may just plain, cussed grower partiality. You see, it isn't just hay we're talking here! I don't pretend to know for sure or mean to promote any brand. Alfalfa was just the crop my father chose and taught me to tend.

Protein and Nutrition by Type of Fodder

	Yield Per Acre In Tons	*Dry Matter In Lbs.*	*Digestible Protein In Lbs.*	*Total Dig. Nutrients In Lbs.*
Alfalfa hay	2.04	3,688	432	2,052
Clover hay	1.48	2,601	207	1,536
Timothy hay	1.23	2,173	90	1,167

Medicago Sativa - Alfalfa

Phleum Pratense - Timothy

Trifolium Pratense - Red Clover

Give fools their gold, and knaves their power.
Let fortune's bubbles rise and fall;
Who sows a field, or trains a flower;
Or plants a tree, is more than all.
A Song of Harvest
John Greenleaf Whittier

Metabolism and Net Energy

"Go feed the cows, son. Can't you hear them bellowing?" Never thinking more about it, other than the fact that they were hungry, I would go toss some hay in the manger affronting the animals. They calmed, chewed for a while, reclined and chewed some more. Later we milked them. That was the routine.

There are so many factors involved in farming – even after a crop is raised and harvested. Think about it. I had no idea about mastication, metabolism, 'net energy' or how one kind of hay was different from another. Someone calculated that a dairy cow makes 41,000 jaw movements a day chewing food. This is split between about 6 hours a day eating and 8 hours in rumination or 'chewing cud'. All that chewing takes a lot of energy. That makes sense, I guess.

The useful energy left to produce growth, milk, fat, wool or do practical work is what is left over after an animal uses what it must to digest food, wander around the pasture and handle all the other necessary functions of maintaining its own body. The calculations depend further on weather, animal health, feed quality and hygiene.

Sorry Dad,

On one extremely cold and black winter night, when the Mini-Cassia wind was curling wavelike drifts across all of Southern Idaho, Dad asked me to do the feeding. He knew I preferred most any chore to hand milking. With his head tucked into a Holstein's flank and warmly huddled over a pail of steamy milk, he was probably secretly smirking over getting the best of me. "Sure thing," I blurted, escaping into the hostile air I had forgotten about.

Climbing to near stack height – a new butt was started only a week or so ago – on an icy ladder, the nor'wester was directed right into my bamboozled spirit. The hay knife was uselessly buried in crusty snow. And the fork tines skated rather than pierced. I began to whine and curse. "How am I supposed to cut this frozen mess?" "It's too damned cold for cows to eat anyway." "Why me?" "How does he expect me to do this?" Whine – Slip – Curse – Rip off a shaving. At this rate the cows would perish from starvation long before the cold got them. My complaining was hidden by the howling wind. Or so I thought.

"Let it be. I'll do it myself." Dad's calm, forceful, condescending retort cut through the bitter darkness. He never said much – just enough with always-stinging precision.

Embarrassed horribly beyond my self-imposed distress, I instantly discovered hitherto unknown strength to quickly melt the perceived impossibilities. Alfalfa flew.

Winds calmed. A new moon rose. A re-sharpened hay knife cut soft and deep. Forkfulls filled the manger to overflowing.

I never spoke to Dad about this miracle. He maintained his cagey stillness as well.

Plus there are differences between animals. Fast-growing, 250 pound beef cattle may burn 4 to 5 therms of net energy per day, while a 1,500 pound horse doing hard work can use up 16 to 20 therms. The same horse on a day off can get by on 7 to 10 therms of net energy. In getting back to hay and why it is important to raise the best kind, take a look at the following table showing the amount of energy contained in one hundred pounds of hay. One hundred pounds is enough to feed a cow for quite awhile – say 7 to 9 days – according to a twelve-month test sponsored by the Oklahoma State University Cooperative Extension Service.

Kind of Dry Roughage	Net Energy Per 100 LBS in Therms
Alfalfa of high quality leaves	44.3
Clover hay, red, high quality	44.5
Timothy hay, early bloom	41.3
Bermuda hay	32.3

One therm is equal to 1,000 calories or about twice the number a fashion model consumes in one day.

It looks pretty even between the top three based on pure thermal energy, doesn't it? Now factor in production. Refer back to the earlier table showing the quantities of dry matter and digestible nutrients produced per acre by each of these primary hay varieties. Given the higher yield of alfalfa and therefore the larger amounts of dry matter and nutrients produced, it appears that the net energy advantage goes to alfalfa. What is less clear is how anyone can figure this all out.

Life on the Great Plains

Other Little Known Facts About Hay

According to the US Census, 11,515,811 acres of alfalfa were grown in 1929. [Since this is about old time haying, I decided to use old time statistics] The average yield was 2.04 tons per acre or a total production of 23,493,505 tons of hay. Back in the good olde days many of those tons were loaded on slow moving, horse drawn wagons with three or four tined pitchforks. [For this puzzle ignore the fact that buckrakes handled a lot of the production] This demanded a great many workers compared to today's highly automated harvesting. To get a rough idea of how many people were involved, assume that an average shock of hay weighed 10 pounds and that a worker could pitch six shocks onto a wagon every minute throughout an 8-hour day. Using the 1929 alfalfa production and a few calculations, an amazing commitment of labor becomes apparent. There are 43,560 square feet in an acre. Imagine this footage arrayed as a very long, narrow field 24 feet wide and 1,815 feet long. That amounts to 4, six-foot mower swaths long enough to equal the one-acre strip. Raking the four swaths into two windrows – one for each side of the wagon – sets up our hay pitching case. [2 windrows x 12' width x 1,815 = 43,560 sq. feet]

The 1929 production averaged 4,080 pounds per acre. Dividing that into two rows results in 204 ten-pound shocks on each side or one shock every 8.9 feet. Does that seem about right to you old time pitchers? How about walking and loading 6 shocks a minute? Reasonable?

If so, in one day, our hypothetical worker could load   →   14.4 tons
[6 shocks x 60 minutes x 8 hours x 10 lbs / 2,000]

Therefore, clearing the fields in a summer harvest season of 10 weeks would require   →   23,307 workers
[23,493,505 tons / (10 weeks x 7 days x 14.4 tons]

And that's just the pitching crew. Add wagon drivers, hay yard stackers, derrick horse drivers, water boys and you have a real setting for dinner or lunch as we now call a midday meal.

One last esoteric point about feeding hay. A great deal of study was done on how the net energy value of hay might be affected by the manner in which it was harvested and stored. The underlying question had to do with whether or not grinding, cutting and cooking feed sufficiently increased the value of the crop to warrant the added expense of the extra labor.

Dinner Time

Remember the dinner bell? Unhook the horses and loose them to the water trough. Slap the dust off your Levis. Wash up in copper tubs scattered on benches in the back yard. Grab a plate. Get in line. The women folk were busy – as busy as the crew – all morning fixing up a meal worthy of royalty and bountiful enough for an army.

Roast beef. Jacket potatoes. Cabbage slaw. Spanish rice. Leafy greens. Milk. Brown gravy. Corn-on-the-cob. Ham. Sour dough biscuits. Green beans. Baked beans. Fresh sliced tomatoes. Lemonade. Coffee. Apple pie. Ginger snap cookies. A Prince Albert roll-your-own. And a 20-minute snooze on the lawn before rehitching the teams.

One research paper concluded that grinding or rolling hay, particularly coarse, stemmed hay, saved the animal much labor in chewing thereby increasing the net energy going into body growth or useful work. Another study contradicted this conclusion, however, deeming it ludicrous. "Muscles do not grow strong through idleness. Exercising the organs of mastication and digestion are necessary for good health and strength".

So should we stop grinding? Not necessarily. Crusher harvesters are being produced and used regularly today. The reason, though, has more to do with shortening drying time than chewing time. A cynical reviewer of this data may wonder if such laboratory research was intended to aid the animal or the manufacturer of new product lines. What say you John Deere? John Farmer? Seriously, agricultural science has endowed the world with tremendous benefits for both sides. That much I do remember from my one year of high school FFA.

Even so, guessing that you, analogous to my city-born wife, can proceed affably without any further background on the subtleties of hay, I will advance to field preparation and seeding. *You're welcome!*

But before I move on, a question:

Did you know the folks were worrying about all this when planning next year's crop rotation?

Photos of the stockmen on this page are courtesy of the Library of Congress, American Memory Series

CHAPTER 2

Making Hay

Climate and Soil

Alfalfa is a robust crop with a remarkable adaptability to climate and soil. It requires considerable moisture to produce profitable yields yet doesn't favor humidity. A relatively dry atmosphere where water is available for

irrigation is best. Alfalfa grows from sea level to the 8,000 foot plateaus of Colorado. Heat is usually not a problem but cold weather, especially in poorly drained areas with alternating freezing and thawing, can destroy plants or cause serious damage.

Well planned and prepared seedbeds are vital for the tender nature of young alfalfa plants. A combination of plowing, disking and harrowing should develop a planting surface loose to a depth of at least 2 inches. Purists run the field with a culti-packer prior to or after seeding to give a perfect terra-cotta firmness.

Soil pH is next on the critical path to success. No other common forage requires so much

Farmers are a quietly prideful lot.

Their action, dress and working style are seldom staged for grand performance.

Yet always there is presence, a unique flair for costume, vivid settings and undeniable demeanor.

It is just a matter of timing and available resources as to how any production comes off.

lime as does alfalfa. Lime is needed both to neutralize the soil and for the use of the plant itself. If the pH level is unknown, a reasonably accurate home test can be performed using red and blue litmus papers purchased from the local drug store. Ball up a moist handful of dirt and touch it to the two litmus papers with care not to dampen the papers with your own perspiration. If the red paper turns blue and the blue paper does not change color, it can safely be assumed that lime is not needed. If the blue paper turns red, however, you should add lime. A formal reading of 6.5 pH is considered minimal for best results.

Because alfalfa is a heavy feeder, it requires an abundance of plant food. Good barnyard manure is an excellent source of humus and other needed foods. Some potash and nitrogen, depending on soil conditions, may be added by spreading commercial fertilizer. Assuming reasonable attention to the foregoing and ignoring a rare geographic concern over proper bacteria for soil inoculation (an overly acidic or deficient vegetable matter condition) the

planting bed is ready for sowing. [This may be a good example of where loess is more, but I digress.]

Sowing

Planting seed is a multi-method, dealer's choice sort of operation, according to U.S. Department of Agriculture Farmers' Bulletin No. 1283. The 1922 publication said sowing "may be done with a grain drill with a seeder attachment or with an alfalfa drill, or the seed may be broadcasted with a hand seeder, a wheelbarrow seeder, or by hand and covered with a light harrow, a weeder, or a brush drag." In rare situations, it can be broadcast onto a covering of snow. As the snow melts it carries the seed into the upper crust of soil, saturating and softening the seed at the same time. Planting is usually done in early spring or late fall so that new plants can get established while the soil is moist and before extremes of hot or cold weather set in.

Early century publications recommended that drilled seeding should be to a depth of about 1 inch in heavy soil – 1 ½ in sand. Today ¼" to ½" is considered adequate. Similarly, the rate of seeding has changed over the years. Where 20 to 25 pounds of seed per acre was once the rule of thumb, 15 to 20 pounds now seems to be the more common recommendation. Again, a lot depends on the moisture content of the land, its fertility and its geographic location. Since the seed is relatively inexpensive, it is wise to err on the high side in order to get a sufficient number of plants per acre.

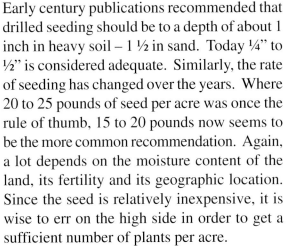

Months are needed to start a hay crop. To compensate, farmers often plant the seed along with some wheat or barley as a nurse crop. In that way, they realize some return from the land while the crop is becoming established. This is more an economic decision than a horticultural one. Without a serious erosion problem, there is no underlying reason for a supporting nurse crop. If used, it is suggested that the nurse crop be removed as soon as possible to permit better growing conditions for the primary crop. Now it is a matter of sitting back and waiting. Sure!

Were we directed from Washington when to sow and when to reap, we should soon want bread.
Thomas Jefferson 1743-1826 3rd US President

Or maybe cultivating, irrigating, cleaning ditches and corrals, fixing equipment, milking, collecting eggs, mowing the lawn, chasing stock, fencing, doing the books and negotiating with the bank for next year's credit line.

Making Hay

Eventually, with luck and technique, a 3-ton per acre field of forage will soon stand gloriously in bloom. In the case of Alfalfa, it will be about 2 ½ feet tall and have tiny pink or blue-violet flowers. When they appear, at precisely 43% bloom, it is time to cut the first crop. But here again it depends on one more factor – that is, who is going to eat the hay. Some studies show cows like it cut a little earlier – say when it is just starting to bloom. Horses, on the other hand, prefer a later cutting – a little too laxative for them when it's cut early.

These may be overly homespun guidelines for, as Mr. Ariel Bean of La Grande, Oregon recalls, they cut at about ten percent bloom and never had any complaints from any of their stock, whether cow, horse, sheep or goat. More of his keen thoughts on making hay will be found in the appendix.

Not to diminish the value of erudite technical guidelines, the decision regarding when to cut,

is usually controlled by fate – such as if the mower is running, the ditch rider says it's your turn for the water or the school principal called about junior.

George Ford Morris   1873 - 1960
Famed American painter of prize winning horses.

Once cut, the real game begins. From here on the factor supreme – moisture – takes control. An old saying sums it up: *"The moisture without won't hurt you, but the moisture within will."* Another adage, possibly more familiar, advises: *"Make hay while the sun shines."* Both relate to the essence of *'making hay'*. For hay is not what is planted and grown in the field. Hay is what is *'made'* in the curing process. It is the fodder resulting from the curing or drying of the plant after it is cut.

Drying hay to 20 or 25 percent moisture content is the goal. Keeping the leaves intact during the process is equally important for it is the leaves that contain two-thirds of the feeding value of the plant. A good farmer will cut the hay one day, rake it in the early morning when there is some dew to protect against leaf loss and

stack it several days later. A lucky farmer times these operations to coincide with a high-pressure heat wave with no rain or wind in sight, and then has it stacked before the weekend or the 4th of July holiday break. A new crop will be ready in 30 to 40 days. Saving some luck for the second and third cuttings is always a good investment.

Hay-topped houses were stacked in some areas when times were tough. Those who tried sod had better results. This one doubtfully ever served as a real homestead. I sense some country tongue in cheek at work. Note the flags on top.

The seriousness of the hay-making process cannot be over emphasized. F. B. Morrison, Professor of Animal Husbandry, Cornell University, warned in his 1947 *Twentieth Edition of Feeds and Feeding:* "Many millions of dollars are lost each year by farmers through lack of sufficient care in hay-making and also because they do not understand certain of the principles involved in the process." He continued. "High-quality hay is leafy; it is not made from plants cut at too late a stage of maturity; it has but little foreign material, such as weeds and stubble; it is free from mustiness or mold; and it has an attractive fragrance typical of the particular crop from which it is made. Such hay is much more nutritious and also more palatable than that which is deficient in these characteristics. Indeed, there may be much more difference in feeding value between good and poor hay of any particular kind than there is between the different kinds of hay made from the common hay crops."

I found this striking scene in a random search one day. Every subsequent attempt failed to find the artist. Hopefully his or her gracious permission to display the work will be as forthcoming as was the talent that created the original masterpiece.

21

Farm policy, although it's complex, can be explained. What it can't be is believed. No cheating spouse, no teen with a wrecked family car, no mayor of Washington, D.C., videotaped in flagrante delicto has ever come up with anything as farfetched as U.S. farm policy.

P. J. O'Rourke, U.S. Journalist   "How to Tell Your Ass From This Particular Hole in the Ground"

CHAPTER 3

Sickles, Scythes And Mowers

The first productive implement used for cutting grass or grain was the sickle. Early writings mention Egyptian sickles with curved blades made of both iron and bronze. Records from Switzerland, dated even earlier, discuss a curved implement attached to a handle that was swung in a partial arc or circle. The Greeks and Romans aided in the development. Pliny, The Elder, a Roman scholar and naturalist, AD 23-79, distinguished between sickles and the longer-handled variation labeled a scythe. He called the shorter one the Italian sickle and the longer one, with two hand holds, the Gallic sickle. In a related piece, Pliny observed that: *"The master's eye is the best fertilizer."* You can see his interest in agriculture extended over a broad spectrum.

Sickles - Then and Now.

After Roman times, for some reason, agricultural development declined for about ten centuries according to a U.S. Dept. of Agriculture compilation by W.R. Humphries and R.B. Gray. They relate that Crescenzio, a noted scientist of the time, described these two implements in 1548 as still having a design similar to those of centuries past. The ubiquitous implements retained their popularity well into the twentieth century.

Even the advent of mowing machines failed to end or significantly curtail the use of sickles and scythes. An 1860 U.S. Census publication stated: "Tedious and laborious as its use appears, compared with that of the mowing machine, it is wonderfully effective in comparison with the crude practice of the Mexican of our day who cuts his grain and hay by handfuls with a common knife." More oddly, in a report of the Commissioner of Agriculture for 1869, a correspondent writing from Marion County, Florida said: "They mow grass here with a hoe. It's a fact. A well-known physician told me the other day that he believed there was not a scythe in the city except mine. Those who don't mow with the hoe do so with a reaping hook." Without dismissing the ethnic slights, it is evident these implements served the popular and practical needs of users over many centuries. Albeit slowly, advancements were accepted in time.

Cradle scythes such as this five finger model not only cut but shocked the harvest when directed by an expert. Picture is courtesy of Richard Van Vleck, American Artifacts Corp.

The attachment of cradles to the snath (handle) of a scythe was a noteworthy enhancement. This European innovation dates to the 13th century. America stalled its adoption until late in the 1700s worrying over seed loss. Though it could be used with less stooping, the roughshod stroking process was considered to cause excessive loss of grain heads – a drawback apparently irreconcilable with the reduction in back spasms.

Deftness in use of these implements is all but lost today. Few in the modern world even possess one – except as antique wall hangings –

and fewer still can exhibit merit in their use. Happily, some cling valiantly to the past in a romantic sense. Peter Vido, in an interesting article for the *Small Farmer's Journal,* lays out, in fine detail, how our cousins of old rhythmically wielded this remarkable tool through the pre-power ages. Excerpts of his article may be found in the appendix including a matter-of-fact treatment of how to adjust your nibs.

Powered Cutting

Modernization struck earnestly in the early 1800s. Inventors in England and America were working on horse-powered mowing and reaping machines. The first recorded American patent was issued in 1803 to two New Jersey men for a 3-wheeled affair that was intended for both grass and grain. Improvements continued for several decades, until in 1840 or 1850 when single purpose mowers – not combination mower/reaper machines – began to draw serious interest.

Recognition is owed to many early day mechanics that dreamed up ingenious ways to make the conversion from hand to power cutting methods. Consider the thought process of Jeremiah Bailey who succeeded in 1822 in cutting 10 acres of grass in one day. His machine consisted of a series of scythes mounted horizontally on a circular framework which in turn was connected to gears in the left wheel. Whirling blades laid down the grass perhaps in sync with the spinning of his own mental cogs. The device even automatically sharpened itself using a whetstone affixed to the frame thereby honing a keen surface with each rotation.

Mr. Bailey qualified for a patent but probably didn't retire off of his efforts with the Bailey Mower. A cohort, however, may have. Obed Hussey, in 1833, developed the reciprocating knife and slotted guard cutter bar that is universally used to this day.

Utah/Idaho border near Manila

Horse Fly
Tabanus Americanus
Remember this little
troublemaker?

After Scything

In the heat of day,
farmers and farm hands

Lift fork loads of clover
in rhythm to the sound of bees
that hums in their ears like the future
of the horse drawn wagon whose dust
echoes the coming of machines. Still

in this summer, **1949,** my father is
working the top of the hayrick,
his muscle and bone, sweet as the smell,
strong as the hone of bee. Caught
in this motion he doesn't feel himself
fall until he sees the ground and the set
aside hay fork sticking out of his thigh.

Then the man, who drives the wagon,
the one who pulls the fork from his leg,
takes his chew of tobacco and slaps it
on the wound, saying this will cure
anything and it does; my father
remembers those days of scything
his youth, like mine, when we were all
certain of cures for everything.

From Field Stones by Robert Kinsley

Literally hundreds of models and patents were issued to enterprising developers over the next seventy years. A major hurdle was crossed in the 1850s when, at test trials held in Springfield, Ohio, the combined mower and reaper machines of Cyrus McCormick, A.J. Purviance and a Mr. Smith (probably of England) competed against the single purpose, mowing only devices entered by Obed Hussey, a Mr. Castle and William F. Ketchum. The battle over cutting supremacy waged on for years. One should understand here that the majority felt it made little difference what was being cut – hay, grain, grass or weeds. A sound cutter bar of some sort was all that was necessary. The trailing platform of a grain reaper had little to do with the cutting that took place up front. This proved not to be the case.

Technology of the time made it difficult to hold the cutter bar close to the ground. This is a major objective in getting the most out of hay cutting. A higher cut in grain fields didn't

matter as much and was actually better for harvesting wheat or oats since a short shaft below the grain head meant less chaff to clutter the threshing chamber. Those favoring the combination style equipment focused properly on the critical platform and binding aspects of their machines. In doing so they lost the cutting war. Depth control mattered.

Single purpose mowing machine advocates focused on the question of how to control the cutter bar as it traversed on uneven and sometimes rocky terrain. They concentrated on drive wheels, slotted guards and rigid versus flexible form cutter bars. Should small wheels be placed under the bar to guide it over the field? How far apart should the fingered guards be placed? How does driver weight on a rearward seat affect control? Could higher torque gears eliminate backing up for new cutting starts? One patent, issued to James A. Saxton, claimed:

> ➤ *...attaching the cutter bar to the frame of any mowing machine using his double hinged arrangement was especially worthy, as the guards, or fingers, or that part to which they are sustained and supported, are free to rise or fall bodily and also to have a lateral or wobbling motion to enable the cutting apparatus to conform freely to the undulation of the ground over which it is drawn, independent of the up and down motion of the mainframe.*

Wamic, Oregon

An article in *Implement and Tractor* stated:

➢ *The mower had its genesis in the same experimental period which produced the reaper, one contributing to the other, as the original functions of the two machines were quite similar. The reaper developed through many ramifications into the binder, header and more recently the combine. [but]* *"The mower is still a mower."*

Could it be the author had a bias? Who knows? The results are what counted.

Machine Performance Trials – 1850s

A number of competitive trials were held in the 1850s to compare and test the merits of the many equipment designs being brought to market. New York and Ohio were the crowd's favorite sponsors of the challenging events. These were serious affairs that doubled as glamorous social events catering to agriculture – forerunners to today's equipment and farm supply conventions.

At the 1852 Ohio trials and others held later in Geneva and Syracuse, New York, single purpose mowing machines were consistent winners. Ketchum's machines were the most frequent winners, particularly after he adopted the Hussey type of rigid bar cutter. Mr. Ketchum is sometimes spoken of as the father of the mower trade – his reputation hinging upon his being the first to put a reliable production machine on the market that was distinct from the reaper.

Peculiarly, although Ketchum's stand-alone mower proved a consistent top finisher, his initial cutter bar didn't. It was an endless chain of knives that simply could not match Hussey's rigid bar invention employing reciprocating knives and slotted guards. Borrowing ideas

and advancing on the merits of others has long been a favored path to progress, it seems.

The Buckeye mower manufactured in Akron, Ohio, achieved the highest popularity rating of the day. However, it was not patented until 1856, many years after Hussey, McCormick and Ketchum made their discoveries and obtained the first patents.

By 1866, things had improved so much that *"every entry was able to stop in the grass and start again without backing to get some advance wheel speed, without any difficulty and without leaving any perceptible ridge to mark the place where it occurred."* This was of particular merit since the drive power was

27

provided by the movement of gears activated by a wheel supporting the mower's frame – not from a yet-to-be invented power take off drive projecting from the rear of a tractor. In other words, the cutting knives could not move or begin to cut until the mower's wheels themselves were pulled into motion. This was radical, coordinated gear meshing!

Other Discoveries

Aside from advances in cutting efficiency, there were many other discoveries in the 1800s. For example:

➢ Continuous oil baths were made possible by using rawhide to form dust proof, airtight gear housings.
➢ Weight distribution was improved by placing the operator seat behind the main axle to reduce neck weight on horses while providing sufficient balance to prevent tongue whipping.
➢ Guards, ledger plates and shoes were streamlined to reduce the resistance of hay moving across the sickle bar.

And cast iron was definitely coming into vogue for, as one manufacturer stated:

"Cast-iron is far superior to wrought-iron for our purpose, as the latter naturally lacks the rigidity of the former, which is absolutely necessary to hold the shafting in perfect line. A machine in actual use is ten times as liable to be put to a light strain sufficient to bend a wrought-iron frame, throwing shafting out of line and making the gearing run hard, as it is to be put to the enormous strain required to break the cast-iron frame of the Buckeye."

These may seem insignificant today but they were decidedly monumental at the time to both the developer conceiving the idea and the operator struggling to cut big acreage with no 'standing misses'. The effort was worthwhile.

And it sure beat swinging an infernal scythe.

Things were in place to cut some real hay with horsepower. There would still be issues – maintenance, sharpening, oiling, greasing, sickle teeth replacement, guard tightening, harness repair, and horse care – but cutting twenty acres a day became something one could really crow about.

Starting a New Field

The best part of mowing was starting a new field. This was especially true for fields lined

Howard Likes Horses

by fences, crisscrossed by irrigation dikes or bound by leaky ditches. The terrain laid hidden in a tall, thick mat of alfalfa which obscured mud and evil sickle killers lurking in wait. The first pass was made opposite the direction that would be traversed during the rest of the cutting. It set the pathway. The sickle end traced the edge of the field nudging but not impacting obstacles. Horses trod on the standing crop for that pass and were steered in slow and intentionally jerky moves to guide the cutter bar as close to the surrounding fence or ditch bank as possible. *No stem shall go uncut* was the unspoken mantra.

Getting stuck or chewing up a fence post was an inexcusable offense. For those who may not have been there, it might be looked upon as one of those macho farm boy things.

The first round took time and patience. The second was faster but had problems of its own. Traveling backwards against the trampled flow placed the cutter bar on, or close to, the fallen bounty of the initial pass. Heavy green clumps plugged the cutter teeth. Simply backing up and restarting could usually clear the blockage. If you had to get off and clear the foul by hand, you did so with extreme care to secure the horses or disable the sickle [here I am referring to sickles powered by a tractor's power takeoff shaft]. Failure to take this caution caused more than a few careless types to tolerate short-fingered futures.

After completing the two setup passes, mowing was mostly a heartwarming ride in an agri-park setting, polishing a tan and inhaling the sweet, sugar-based aroma of freshly-cut lucern wafting in the clear air of a gentle summer.

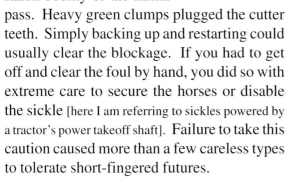

I used to strap a twenty-two rifle to the mower frame. Pheasants and rabbits made their homes in the lush cover and were occasionally irreparably injured when they froze in the face of impending danger.

Width of Cut and its Importance

From the 1881 New Buckeye Mower Co. files:

Let two machines, one cutting four feet wide and the other five, follow each other in grain [or alfalfa] *for one day, each being drawn a distance of 24 miles, the former will have cut 12, the latter 15 acres. To cut 100 acres, the former must travel 200 miles, the latter 160 - a saving of 40 miles travel with a heavy load, which is a matter of no small importance.*

It is a saving, too, of 40 miles walking, or 10 miles for each of the four binders. It is a saving also to the farmer of two days' labor for himself and team, and the wages and board of the four binders for two days in cutting 100 acres of grain. The Buckeye Table-Rake with ordinary driving, will cut one foot wider than any reel-rake having same length of bar.

Excerpts From New Buckeye Advertising Booklet
Farm Equipment Manufactured by Aultman, Miller & Co., Akron, Ohio
Circa 1881

The merciful man is merciful to his beast and will keep his Buckeye Mower knives in good order and journals well oiled.

A perfect godsend to the weary housewife. Saves time, labor and expense, both in field and kitchen is what they all say of the Buckeye Binder.

You cannot fatten your hogs on scatterings left by the Buckeye Binder; much less will it pay to glean your fields after it.

The Buckeye Down Binder is big enough to do the work, light enough for two horses and will pass through an ordinary farm gate.

Your boy may safely run the Buckeye Mower and Reaper, as the driver's seat is at a safe distance behind the knives.

As opposed to, Buckeye recounted: A son of Adam Schiesman, living North of Carroll, Iowa, was thrown from the seat of a Champion Rear Cut Reaper, and cut nearly in two and killed.

New Buckeye Front-Cut Mower, Manufactured by Aultman, Miller & Co., Akron, O.

CHAPTER 4

Rakes And Tedders

Gathering Hay

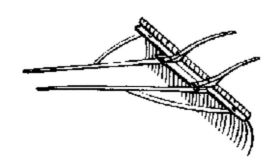

Hay, as mentioned previously, is what is made in the curing process. It is the fodder resulting from the curing or drying of the plant after it is cut. Raking and mounding it loosely in cocks or windrows to air dry slowly is an essential step in the curing process.

In the really good olde days, scythe cut forage crops were gathered with a hand rake and cocked to cure with wooden forks. Hand raking was a relaxing, pastoral task. It gave rise to many of the bucolic panoramas we now cherish. Normally women and children performed this toil while men tended more to scything and stacking. Most tools were hand made.

Merry Old England

Forks were carved or cut from a single shaft of hickory or similar hardwood. Wedges spread and held the shaped tines. Two extra tines, on this model, were formed and attached for extra volume.

One historical source recounts that the first break in the all-hand-regime came with the 'Whoa-Back' rake. Dating to around 1840, this horse-drawn discovery collected cuttings from each swath and pulled the mass in line with the deposited collections of the previous round. The operator then yelled whoa and backed up to release the hay, forming an ever-lengthening windrow. Whoa Backs resembled the popular dump or barrel rake but were built without the wheel-mounted ratchets invented later that would aid in lifting the tines. Previously, dumping was done laboriously by physically heaving a handle with assistance at times of a foot lever.

This painting, the hay field scene on the preceding page and the black and white sketches of antique rakes are courtesy of the Rural History Centre, University of Reading, UK.

Virtually all of the first-to-market inventions were single horse units. The rake was hooked to the horse by two long shafted thills similar to those used on racing sulkies.

In spite of the claims of the Whoa Back being first of its kind to market, another short-lived horse rake is dated to 1820. History is admittedly unclear and even contradictory according to the authorities found in my research. Whether it preceded the wheeled rake or was introduced simultaneously is open to debate. Nevertheless, a counter point to the Whoa Back was the wacky 'Flop Over' – a very interesting device indeed. It consisted of 15 to 18 wooden teeth projecting from both sides of a headpiece or center rail. Ropes were attached to the headpiece and hitched to a horse or ox. Designed as a non-riding walk behind, this nonstop machine was capable of handling 2 to 3 acres per hour. In use, the teeth ran flat along the ground, passing under and collecting either swath-cut or windrowed hay. Handles projecting to the rear were used to steer the rake and trip the load when full. When the handle was lifted, the tips of the wooden teeth caught the ground causing a half revolution or flop over of the rake. It then glided over the dumped load and set up the opposing tines for a new load all without stopping. The operator loped along behind at a quick pace, deftly throwing the handle every few feet while sweating in glee at his unrivaled production!

Bragging on the merits of another entry, an avid promoter testified: *"They throw the windrow into heaps or bundles of 80 to 100 pounds each, ready for cocking or loading. A boy and horse can thus rake and bunch 20 acres a day."*

Times Change!

What do you suppose would happen in today's world if an equipment manufacturer issued pronouncements influencing children toward achievements of this nature?

Barrel rakes can still be a barrel of fun! Here, my mother, brothers, sister, aunts, uncles and cousins celebrate a whacky spontaneous reunion. I swear it was not a setup photo op for this book.

Old Macdonald had a farm,
E-I-E-I-O
And on his farm he had a cow,
E-I-E-I-O
With a "moo-moo" here and a "moo-moo" there
Here a "moo" there a "moo"
Everywhere a "moo-moo"
Old Macdonald had a farm,
E-I-E-I-O

The US Patent office evidenced impressive inventor activity during the nineteenth century. A summary of new issuances included:

1822	the Flop Over
1839	the Spring Tooth
1848	the Dumping Sulky
1850	the Draft Dumper
1852	the Self-Dumper;
1856	the Spring Tooth Self-Dumper
More	Draft Dumpers in 1856 - 1884
And	Drag Dumpers in 1866 - 1870

The US Commissioner of Agriculture noted that seventy-four patents for horse hay rakes were issued in 1870 alone.

Curiously, mechanized raking preceded mowing. The Commissioner of Agriculture reported in 1872 that: *"The horse hay rake was invented at an earlier date than the mowing machine. It has been used in this country nearly seventy years, and the saving by its use, sixty years ago, was estimated to be the labor of six men in the same time."* I am not sure I understand him but the quote is accurately presented.

A rake by any other name:

Hand rake, revolving, flop over, spring tooth, dump, barrel and sulky are names covering the progression of devices used in collecting hay from a freshly mown swath.

Scatter rakes were 30' wide versions made by welding dump rake bodies together to gather missed cuttings from recently harvested fields.

Buck rakes, also called sweep rakes, push rakes and 'go devils' were really not rakes as such. Instead they had long wooden forward protruding teeth that were used to pick up previously raked windrows and transport the loads to the stacking area. More on these later.

My Team. A poor photo but great horses.

I returned from my freshman year of college to spend the summer working for Wesley Rogers, a nearby neighbor. During much of that summer I contemplated the south end of this aging northbound team.

Though not lucrative, the experience was enjoyable and did much to rejuvenate my scholastic aspirations. The rhythmic clink, thunk, stomp and clang harmonized with endless cyclic foot stomps, pawl gear locking into wheel hubs, hand lever whipping and the thudding drop of tines setting for a new drag.

Step Clink Pawl Thunk
Stomp Thrust Clang

Step Clink Pawl Thunk
Stomp Thrust Clang

Step Clink Pawl Thunk
Stomp Thrust Clang

On long rows, the horses traced the windrows, flexed muscles memorized the routine and my mind sleepwalked among the growing shocks. All was dreamy until the familiar cacophony changed suddenly when I struck a rock, a fence post or I dozed off and ran into a ditch.

As seen in the accompanying photo, one worker [*probably an aspiring college student*] is seated on a barrel rake dragging freshly cut swaths into a windrow. This was a common use of dump rakes, i.e. dragging the tines over the stubble to collect fresh mown hay into long windrows. Foot soldiers followed behind hand-forking the windrows into individual shocks.

The advent of side-delivery rakes all but eliminated dump rakes from the windrowing process. When still employed, dump rakes were relegated to individual shock making. In that use, the driver steered his horses and rake astraddle side rake formed windrows making short drags and tripping the tines frequently to produce shocks mechanically instead of by hand fork. It took two passes down each windrow to complete the process.

The first pass broke the windrow and got a shock started but left a protruding tail of hay. This was caused by the over-passing of the

rake as it rolled down the windrow before the tines could drop and start raking again. In the second pass, made going the same direction as the first, the initial mound was turned over onto the protruding tail and further balled into a cone-shaped cock of proud workmanship worthy of any pitcher's respect. It took precise timing and a touch of flair with the hand lever to do it just right – a macho trick of the dump rake squadron.

Side-Delivery Rakes and Tedders

Side-delivery rakes were mavericks in the mechanized hay-curing world. They came into existence around 1910, doing duty as diagonal rakes whisking swaths into neat windrows more quickly than a barrel raker could ever hope. The long slanted frame supported a moving framework of three or four bars to which were affixed gangs of 6" stiff wire prongs called kick forks. In motion, the bars spun rapidly creating a whirr of short, gentle strokes that rolled the hay sidewise and backward into a smooth tunnellike bun.

At other times, with some adjustments to reverse the machine's rotation, side rakes served as tedder rakes, gently lifting and fluffing the hay for better drying. In times of rain or 'hay devil' wind, the side rake swiftly

American Gothic won national acclaim in 1930.
By Iowa's famous painter Grant Wood 1891 - 1942

remade the damp, gust-damaged columns. Whichever way a side rake was used, its rotating action built up weight at the far end. This virtually mandated left turns at each row end. The asymmetrical design of the machine created heavy torque that could damage the right front wheel. Right turns could be made if done slowly and your dad was out in the back forty but you damned well better not crack a tongue brace or twist a tooth. The oddly raked end of the row pattern would be a dead give away of your unseemly infraction.

Take a break East of Baker, Oregon, to visit a fabulous 20 acre collection of old time equipment. It's on the way to the Oregon Trail Monument.

There was also a trick in piloting a side rake to sweep two mower swaths into a windrow. Several patterns were known to exist. Remember farmers are independent sorts and favor personal methods. The one I learned:

Side Rake Turning Scheme

1. Start a new field raking outbound on swath cut number four. This sets up half a windrow.
2. Turn left at the ditch and return toward the starting end raking swath cut number one.
3. Turn left, as usual, and proceed outbound again, this time on swath number six.
4. Return raking swath three onto half-done row four, finishing the first two-swath windrow.
5. Continue the pattern always starting outbound on the second swath to the right of the last run and returning on the third swath to the left of that outbound pass.

A careful observer will note that swath two is left un-raked using this diagram. That is where a right turn is sometimes handy. Carefully execute one on the last pass as you head back home. Watch out for the old man. Of course, you could make a 270° left turn and avoid the whole troublesome affair.

As to tedders: they were something else. Their primary purpose was not so much to collect hay as to flip partially dried swaths into new loose mound arrangements so that air could circulate freely and hasten the drying process. The side rake, with dual capabilities, led to the decline of the single purpose tedders. However, modern day manufacturers still offer a wide array of tedders that fluff, spread and even windrow. So does that make them a modern side rake? Amazingly, the intricate arm movements of the tedder arms portrayed in this Bullard advertisement are remarkably replicated in many versions of modern day machines.

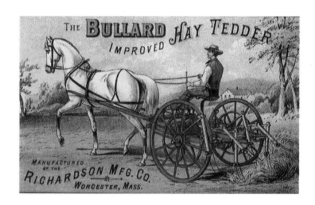

And so it was. With the raking done, Mother Nature took control. Subject to her favor, we could hook up the wagons in a few days.

Who more blest than toil weary folk snug wrapped
atop the last load of summer's harvest?

Scanned from an uncredited Veterans of Foreign Wars greeting card.

The farmer should make an effort to keep his crops clean of weeds, that he may not have to leave the mowing field to cultivate his crops when the grass needs to be cut. The old method of leaving the grass to stand until the seed is nearly ripe is not a good one; it is now very generally conceded that grass should be cut when in full blossom. Under the old system of hand labor, many large farmers could not cut all of the grass at just the right time, but for want of a sufficiently large force of laborers, some of it had to be left uncut until the seed had ripened; but now with improved implements, one man can accomplish in one day what it formerly used to require several days to do with hand labor.

Extract from The Old Farmer's Almanac by Robert B. Thomas. July 1900

CHAPTER 5

Wagons and Loaders

Happy Memories

Some of my first memories are of wasps, bombers and a tipping hay wagon. I think it was in the summer of 1942 or 1943 when I was around five. America had entered WW2 and the B-24 Liberator bombers were training out of Mountain Home Air Base near Boise, Idaho. It was a warm morning and I was pouring dry sand into a nest of ground wasps. The buzzing activity captured my interest. On another trip to the sand box for more fuel, I heard a dull roar. It shook the sky yet there was nothing visible between the scattered clouds. The noise magnified. Giant planes came slowly into view, flying majestically from the West. Planes and more planes. Hundreds of them. In a long, wide pattern. Wings seemed to touch. The roar was magnificent. The sun eclipsed. Why they were there or what they were doing was beyond my

TIMOTHY AND RED CLOVER FIELD NORTH DAKOTA.

young insight but the formation dazzled my mind. Then they were gone. I knew that one day I would fly but for now I went back to wasp work. They did too.

To get my mind off the bee attack, Dad took me out to the hay field. It was to be the last load and everyone walked out for a *'happy ride'*. I remember Aunts Dora and Esther being among the crew. Everyone clambered up the front ladder panel. Someone threw me aboard with a half spin. I recall Dad driving the team home but it wasn't like him to go at a trot diagonally across a dike.

Dikes are long mounded rows of dirt formed by plowing two runs together. They are used in flood irrigation to break the field into smaller sections and keep the water flowing in a restricted zone.

Ririe, Idaho

39

More likely, it was insouciant Uncle Ernie at the reins. At any rate, as the front left wagon wheel crossed the dike followed quickly by the right, the wagon started to rock. The left back wheel hit next and the load went south. All and sundry hit the stubble – covered with hay. The ground was hard. The stubble fresh cut and sharp. But no one got more than a scratch – nothing like the bee welt over my eyebrow. My aunts looked for cuts and bruises and embarrassed me with their mothering. Dad and his brothers asked if I would like to try it again. We laughed, reloaded and clicked the horses to the stack yard. What a day! Bombers, bees and bounces.

Goat powered wagon on Henry Anderson's Farm Milton, North Dakota
North Dakota State University collection

This post card was mailed April 9, 1910 to a Mr. Lawrence Christensen of Bickleton, Washington. His sister hoped he was finishing the haying.

Wagons

Wheeled carts handled the many hauling tasks of early farmers and merchants. Deliveries were pulled or pushed to their destination by their operators. Loads were small and distances short. Quantities were often cut to fill only the needs of a day for a hand full of animals rather than providing for a large herd over a cold winter. In such cases, it made sense to build and use smaller apparatus, especially if they were powered by hand.

The range of wagon styles is a remarkable study in itself. Many were designed and built by gifted owner/users who possibly enjoyed perspectives less demanding than their time pressured descendants. Sizes and styles ran a gamut from stubble hugging slips to hybrid one or two wheeled drags to full size four wheeled wagons. As larger loads became necessary, non-wheeled slips replaced carts. Other farmers turned to long toothed sweeps that glided along the ground pushed by one or more animals. The extended teeth slid under windrowed or cocked piles becoming virtual self-loading wagons.

One of the strangest looking devices had wheels mounted only on its front end. This raised the front of the wagon bed above obstructions and reduced the weight being dragged across the field surface. The rear end slid over the ground with little gouging.

Ironically, the rear wheel bunk is easier to mount than the front since it does not have to pivot when turning. Why craftsmen made the effort to install front wheels and not take the time to install a back end is a minor mystery. It could have been economics. In any event, the number of these contraptions placed in service was never huge. Still, as noted in this picture, they continued in use well into modern tractor farming times. As did their entirely bare bottomed cousins. Wheelless slips were used well into the 1940s.

Pitching

The choice of wagon took some pondering. Loading it was more ponderous. To me though, loading hay with a multi-tined fork was one of farming's most enjoyable tasks. Weather conditions were usually ideal. The work was strenuous but not overly so. If you were in shape and had a feel for leverage, the chore was a rhythmic exercise. A crew of interesting people showed up with new stories and pranks to share. You could work on your abs and sun tan at the same time and even take a quick jump in the canal to cool off after every load or so. Not bad for something called work.

Pitching was the heart of fieldwork. It was graceful 3-part motion similar to waltzing – jab, rock, and lift, jab, rock, and lift. As the fork man approached each shock, he leaned forward jabbing tines in at a 45-degree angle at a point about halfway between the ground and the top of the shock. Body weight pushed the fork into the hay, moving it slightly to break any linkage to the ground. With the load secured, the worker rocked backward on his heels simultaneously lifting the weight by lever action [not a muscular arm lift] to a comfortable point sufficiently high to clear the plane of the load. The backward rock and lifting action was coordinated with the distance to and the roll speed of the wagon. Properly balanced, minimal physical effort was expended and none wasted in setting the load on the wagon.

Joe Ilg moved from Willersbraun, Germany in 1926 at age 25. He worked free for three years to pay for passage. Then Nebraska's drought forced him out. In Oregon, his daughter, Cathy Cameron, recalls her family hand pitching fresh mown hay into cocks to dry. She knew they weighed far more than the 10 pounds I was talking about. Her editing of this book was much appreciated but on this occasion, I referred her to page 44.

Photos above and left are courtesy of:
  Peter Henry Emerson Series
  George Eastman House Archives
  Rochester, NY.   www.GEH.org/ne

The 1919 2nd crop alfalfa shot is another
  Library of Congress, American Memory
  Series contribution.   www.LOC.gov

Which limbs – left or right arm or leg – were extended in the jab depended on a person's pitching style and/or which side of the wagon he preferred to work. Many workers could perform equally well from either side of the wagon. Other pitchers, though skilled in the fluid set and pick movement, doggedly insisted on taking only the left or right side. This choice of favored sides was also evident in other toils such as bucking spuds or cleaning ditches with a shovel. I always thought the ability to work both sides was more a matter of mind control than ambidexterity. But some could adjust. Some could not.

GREETINGS FROM NEBRASKA
"THE CORNHUSKER STATE"

Field loading with pitchforks became standard practice for large harvest operations. Long toothed self-loading sweeps worked well for short hauls but were too time consuming when capacity and distance were factors.

Getting shocks onto the wagon was not the end of pitching. Proper placement saved a lot of time for the wagon rider. In fact, experienced pitchers could pitch and arrange

Hultstrand Family Farm, Fairdale, North Dakota This Library of Congress picture depicts stacking hay totally by pitch fork. Note workers on wagon and hay stack in background.

a load from the ground without the need for a man to work atop the wagon – except for some occasional tromping to set the load. It was a matter of engineering. The first shocks were set closely side-by-side, flat side [ground side] down along both sides of the wagon. This left a small uncovered channel down the middle of the wagon. The next series of shocks were placed topside down in this area so that the outer edges of the center row extended over the outer rows, interlocking the base. Following this general pattern throughout built a stable, straight up and down load which handled well in route and during the unloading process.

Contrary to the view presented in the Nebraska greeting card, a pitcher working standard size shocks would walk and pitch while moving alongside and in the same direction as the wagon. There was no need to pick up and carry the shocks any great distance.

Farm Life near Keating-Summit, Pa.

In the photo with the black and white pooch, the hay had been raked in large mounds called cocks. A wagon was pulled up close and stopped while the pile was pitched aboard.

Note that the load was taking on an angled shape above the sideboards indicating that the load stacker - young boy? - may not have been taking the job seriously. He has a fork in hand but probably was there mainly to drive the horses and do the tromping.

The artistic shot promoting Pennsylvania haying shows a loading scene in reliable overview but suggests more refined load top activity than was usually the case. Field pitchers often loaded and set the shocks from the ground without load top help.

Paul, Idaho

By the way, the shocks shown here probably outweigh the 10-pounders of my puzzle.

Hay Slings and Jackson Forks

There was another important factor in how a wagon was loaded. Logically, it had to do with how the hay would be unloaded. Two methods, aside from simple hand fork unloading, were popular – hay slings and Jackson forks. Hay slings were generally favored for large load operations involving derrick stacking, as discussed in chapter 7. Jackson forks handled smaller loads per lift but could put up a lot of hay and had at least one advantage over chains – placing hay inside tight haymows.

Hay slings were positioned lengthwise over the floor of the wagon with looped connectors extending over the wagon ends for easy location and hooking to lifting pulleys hanging from derrick cable lines. As the load was lifted, the flexible slings curled around the jag of hay and securely deposited it almost unruffled on the stack in one giant unloading pass. Two or more sling settings could be used on heavy-duty wagons. The pitchers would load the first chain with what they estimated would not overburden the derrick boom or barn track unloading system. A second sling was then unfolded across the first sling setting. The entire process is clearly illustrated in a series of pictures on the following pages.

Most commercial slings of this type used manila rope between heavy hardwood cross bars. Widths ran from 4 to 6 feet. The largest model sold by F.E. Myers at the turn of the century was a brand new offering. It had 10-foot cross-sticks and was 16 feet long. Flexible wire cable was substituted for the ordinary rope, which as they stated, *"naturally had three times the strength of rope and was much less liable to wear."*

A Commercial Hay Sling

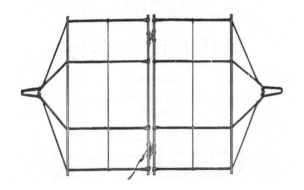

The Famous Jackson Fork

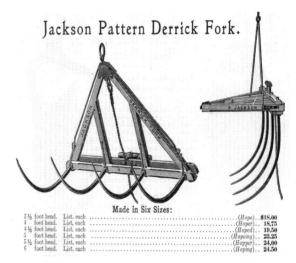

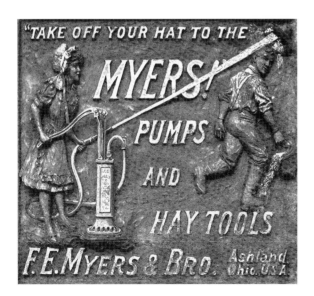

A proud 130-year-old Ashland, Ohio company named *F.E. Myers & Bro.* specialized in two seemingly unrelated product lines – Pumps and Hay Tools. I salivated upon seeing the Hay Tools part of the title in a Reno, Nevada antique store. The publication date is not recorded, as is customary for pricing catalogues, but can be set at around 1906 based on dated quality awards the company displayed in the preface of the 427 page catalogue. The company graciously permitted my extensive borrowing to illustrate early century equipment including superb schematics on how they were used.

All models were adjustable to fit the length of each rancher's wagon. Some used ingenious metal castings to quickly wrap and fit the ropes to desired dimensions. Prices for professionally made hay slings ran from $4.80 for 4 footers to $26.75 for the brand new super duty 16 footer.

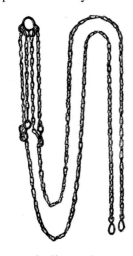

I first learned of commercial slings doing research on this project. On our farm in Southern Idaho, we used galvanized chain to make our own two-strand slings that worked well under careful field loading practices. Two lengths of chain, long enough to span the wagon with a little extra hangover were joined at one end by a large ring. Each length was laid out about 18 inches in from the side of the wagon with the connecting ring hanging over the front wagon panel or ladder. The other ends hung loosely off the back of the wagon so they could be found at hook up time.

At the stack yard the front ring was hooked to a pulley hanging from the master lifting cable. The loose ends dangling from the back were linked to a tripper device riding on the same cable. Once the load was hauled above the stack, it was pushed into position by the stacker. The trip command was then given but unlike the center-splitting hay slings, only the rear chain ends were released as the tripper snapped open. The ringed end remained attached to the sling pulley to pull the doubled chain from under the dropped load and return it to the wagon for the next use.

Tripping mechanisms were highly engineered and much ballyhooed in the catalogue.

Whether affixed to commercial center split nets or two strand galvanized chains, these essential devices depended on intricate locking mechanisms. They had to be reliably strong to hold heavy loads yet flexible and easy to release at the right time. A lightweight rope was tied to the tripper locking pin and held loosely by the ground man until the head stacker ordered him to 'trip it'. The ground man then jerked sharply on the rope, unlatching the snaps which allowed the hay to fall through the splitting slings or chain ends.

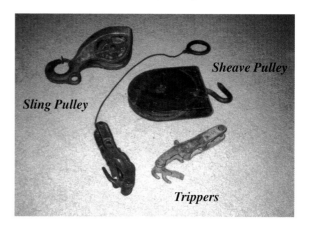

Sling Pulley

Sheave Pulley

Trippers

Piercing Jacksons

Jackson forks required more coordinated labor to set and move a load and were much more dangerous to operate than slings. A minimum of three people were actively involved – one on the wagon to receive and set the fork and work the tripping mechanism, a second to drive the horse pulling the load up the stack and a third working the stack or haymow. Three people commonly manned a sling chain operation as well but the person working the wagon had a relatively effortless job hooking up and tripping the chains once or twice per load. With a Jackson fork he had to set the fork, lock it for tripping, pull it back for another setting and repeat the operation three or four times for each equivalent sling chain lift. The swinging, sharp-tined instrument had to be cautiously lowered by the horse driver and

This is a copy of page 287 of F.E. Myers' 1906 product catalogue. On this one page is captured a bounty of information in engaging turn of the century format. The view may seem slightly skewed. It was in the original as well - the only imperfection in this industrial masterpiece. I found the catalogue in a Reno, Nevada antique store. The original owner/dealer's name was stamped on the inside cover - Mitchell, Lewis & Staver Co. of Portland, Oregon. Both companies are still in business.

carefully attended by the wagon person as he balanced on a spongy wagonload of hay. His sling chain counterpart could do most of the work from solid ground with a fraction of the effort.

Jackson forks were usually purchased, though mechanically able farmers could make their own if they had a good

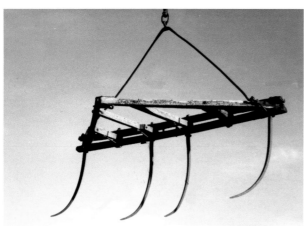

Myers catalogue devoted 5 pages to illustrate 18 models of less well-known rivals known as grapple forks and harpoons.

They, like many other functional and practical tools of the time, enjoyed a commendable history before grudgingly succumbing to progress. Chief among perceived weaknesses of grapples and harpoons was a distinct cumbersomeness in setting and releasing efficient size loads.

forge. Sizes ranged from small 3-½ foot wide, 4 tined models costing $18.00 to five tined, 6-foot monsters costing an extra $6.50. They weighed up to 70 pounds. That much pointy steel swinging from the end of a long cable was more than enough to rivet close attention.

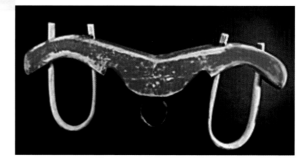

Grapples and Harpoons

Hay slings and Jackson forks garnered the broadest fame and usage. Still, as usual they were preceded by other hay unloading devices, some of which lingered on and competed valiantly for the hay unloading market. The

Another excellent Myers portrayal on the next page traces the contribution of these devices. I have no personal experience with harpoons but have to wonder if they ever consistently handled load sizes suggested by the lower right hand diagram.

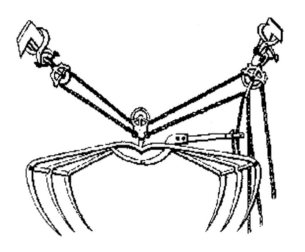

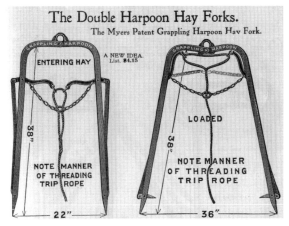

The Double Harpoon Hay Forks.

The Myers Patent Grappling Harpoon Hay Fork.

ENTERING HAY

A NEW IDEA.
List. $4.15

LOADED

38"

NOTE MANNER OF THREADING TRIP ROPE

38"

NOTE MANNER OF THREADING TRIP ROPE

22"

36"

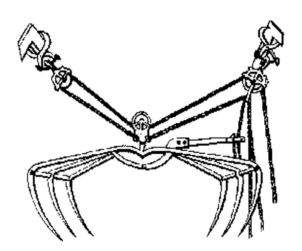

Excellent illustrations showing the transition of hay tools through the years. Note, for example, the long handled version of a Jackson fork at lower left. A short trip rope is built into the mechanism. It was set by the wagon man as he jabbed in the forks. Apparently the head stacker grabbed the handle, set the load where he wanted and then pulled the trip rope himself.

The Myers O. K. Unloader

Arranged to Handle Two Forks

Elevates the Load at Right Angle. Extra Long Truck.

4 FT OF ROPE

TRIP ROPE

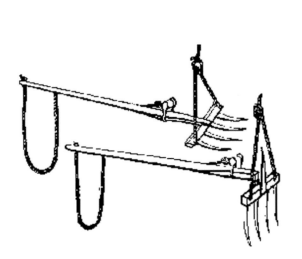

Pitching shocks by hand to clear a field, be it for sling, Jackson fork, harpoon or grapple,  was my favorite position on the hay crew. Folks enjoyed themselves walking the rows and tossing shocks onto a slow moving wagon while joshing with the wagon man who rode aloft, clumsily arranging and tromping down puffy shocks chucked his way from both sides. Even critters liked hay season. Bull snakes loved to lounge under the shocks, away from the hot harvest sun. These mouse eating, rattlesnake-killing farm aides were a harmless lot but like all snakes had that unique power of immediate stupefaction. Pitching them onto the wagon for the utter amusement of the struggling stacker markedly enhanced the passing of hay season.

Then there were runaways. They were rare, but when they happened they launched electrifying episodes of pure anguish. The realization of a runaway about to explode is a never-to-be-duplicated euphoric, heart stopping 'torture-thrill' packed into one split second. These horsefly inspired pandemoniums, reminiscent of chariot races, were inestimable character builders – not just during the feral moments of the frantic race but for days on end in retelling buddies of your equine mastery. Once recovered from the initial impact, if in fact you rode out the opening jolt and got braced for the worst, you morphed gloriously into Charlton Heston of Ben Hur fame. There you were, hunkered down, all alone behind a wild-eyed team. The rest of the crew stood by, watching helplessly in dumbfounded amazement. You set your legs firmly on the wagon floor and magnificently reined for control. The two hell-bent Clydesdales dragged you and the remains of the load across fields, over ditches and through fences right up to the moment they ground themselves to a frazzled stop. Absolutely matchless! No modern day theme park ride can ever compare. And it was free.

My brother provided some chilling, unplanned excitement of his own one warm afternoon. He was driving our tractor pulling a hayrack. Dad and I were pitching. Robert was telling us about little birdies zinging around his head. Neither Dad nor I could see them but we passed his chatter off as a boredom passing game. We were busy. He was playing around the cowling and his babble became zanier but he was maintaining a fairly straight course. That is until he rolled off the right side of the slow moving tractor. Fortunately our old J.I. Case was one with a long steering rod

My friend, Dick Stone, a real life city kid at the reins!
His nomadic childhood placed him in urban capitals such as Yachats, Scotts Mills, Shawmaut, Meeteetse, Withrow and Peerless - possibly accounting for his enthusiastic, rural know-how.

traversing along the outside of the cowling to the front wheels. It deflected his fall sufficiently to permit me to drag him safe from the path of the rear tires. In time we discovered that the fuel cap had been removed. He was gassed. We were flabbergasted.

Loading Technology

Many saw pitching in a less glorious light. Man's consistent endeavor to minimize the toil of daily life produced some exotic machines to eliminate drudgery.

Starting as early as 1848, patents were issued for machines that automatically elevated hay onto wagons. *"It is a machine which saves much hard labor."* stated one patent application. I have to admit the front loader shown here was fascinating and may have saved some sweat getting the load into the crate-shaped wagon. But how about getting it out?

Later patents dating into the 1860s moved the loader to the rear of the wagon and changed the drive mechanism to rise and retreat using a crank gear in the wheels. Two types were popular. One used oscillating rake bars that pulled hay up a slanted ramp where it fell in a cataract onto the rear of the wagon. The wagon driver moved the hay to the front with the ubiquitous pitchfork. Dad bought one for $100 in the late 1940s. It did a pretty good job when the windrow was narrow and straight but

Great scene showing an oscillating bar loader clearing the productive plains of North Dakota.

otherwise left tailings in the field. This irritated my mother to no end. The loader was gone after one season.

Critics found the machine deficient because it shook off too many tender leaves. Some worried if there was enough demand to warrant manufacturing on a large scale. Others complained that it brought hay up too fast overworking the loader who had to fork hay 12-15 feet forward.

The second model used canvas. Attempts were made to replace the moving bars with an endless apron. The apron traveled up and around the slanted ramp and managed to get more leaves onto the wagon but really didn't meet with much more success than the oscillating model. Probably the greatest attraction of these inventions was the elimination of the pitching crew. It was hard to find workers, especially on remote ranches. One or two people could make a round trip in about the same time as a full pitching crew. The machines garnered only modest acceptance. It was not long until a newfangled machine called a baler put an end to the screeching lifter arms. In respect to loose stacking tradition, we shall forego any further mention of those cussed string-popping creations.

51

Sweep Rakes

Improvised variations on ancient self-loading slips resulted in what became known generally as a sweep rake. Countless adaptations and uses introduced widely varied nomenclature to our lexicon including Push Rake, Buck Rake, Booster Buck and Go Devil. These single operator machines were capable of collecting and moving massive amounts of dried cuttings.

Messrs. Humphries and Gray, in their *Partial History of Haying Equipment* wrote in 1949:

My cousin Tony DeBortoli is hidden behind a loaded sweep rake in the above photo taken at his ranch near Rock Springs, Wyoming. Like so many other immigrants to America he settled on a western farm. I wish I had thought to ask what it was like to use a buck rake for the first time. Or if he recalled his parents ever stacking hay on alpine fields using pillow sacks.

"The sweep rake is an implement which has for some time been extensively used on farms in the Middle West and West, and more recently, especially coincidental with farm labor shortages, by the farmers in the eastern states. This type of rake, developed during the First World War, was largely horse-drawn. The sweep rake consists of several long wooden teeth lying almost flat on the ground, pointed at one end and fastened to a strong framework at the other. The point *frequently consists of a steel cap fitted over the end of the tooth and shaped so as to prevent it from running into the ground under ordinary conditions, yet so as to slip under the hay, no* matter *whether lying in piles, windrows or swaths. The teeth usually are about 8 feet long and placed approximately 1 foot apart. This gives a surface of approximately 8 by 12 feet on which the hay can be loaded and carried. A skillful driver with a good team of heavy horses will carry loads of from 800 to 1,000 pounds. Even where comparatively small loads are carried with the sweep it will be found considerably quicker and cheaper than loading on a wagon by hand except where the hay is hauled a long distance. There are a number of types of sweep rakes – one with no wheels, one with two, one with three and one with four wheels. Rakes that had no wheels were sometimes called 'go devils'."*

Dain Alfalfa Power-Lift Rake

Two styles of buck rakes made by Dain Manufacturing Co. of Carrollton, Missouri. The company eventually sold out to competitor John Deere.

Dain Folding Rake

52

Ingenuity reigned supreme as shown in this Scott County, Kansas photo from the files of the Library of Congress. Mounting a sweep cradle to far-reaching rails and gearing the frame and wheels to hoist the load fashioned the *combination sweep rake/stacker.* Notice that the stack is at least 16 feet tall

and the loader still has room to work. Combination rigs could gather loads in the field, transport a half-ton or more to the stack yard and lift it skyward with only one worker at the reins. Usually, though, these special purpose rigs were stationed close to the stack to handle loads delivered to it by standard model sweeps.

Summing Up The Ways

Cured hay laying bare is a highly vulnerable crop. Summer wind and rain, nature's fickle forces, can quickly sack a season's toil. At least, require second and even third rakings to realign and fluff dampened windrows. The costly rework stops molding but knocks off tasty leaves. No wonder farmers experimented with so many methods and machines. Pillow sacks, pony carts, slick bottom slips or drags, and two, three and four wheeled wagons were all part of the haulage litany. And a similar selection of unloading tools – pitchforks, slings, grapples, Jacksons, dual purpose gather and transport sweeps and buck rakes.

Hand filled pillow sacks grace an Alpine field. Spiked, two wheelers also hauled old country hay. The wagon on the right handled the output.

Care for a drink before heading out for another load?

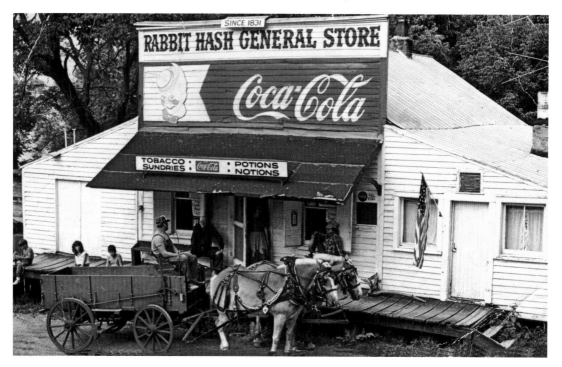

It was a charming era, wasn't it?

All The Fun Of An Old Fashion Hay Ride, Nova Scotia

CHAPTER 6

Hay Mows and Trolley Carriers

Long before irrigation transformed desert dust into productive fodder land, hay was raised in areas with wet winter climates. Cover was essential to protect the collected cuttings of summer's effort. Ground floors in long-ago homes often met that requirement by housing both the feed and the stock that thrived thereon. The patriarch and his extended family occupied the upper levels with usually one family to a floor. In great-granddad's day, the linkage of barns and houses served another now forgotten purpose – heat. There were seldom any interior stoves in our ancestors' homes aside from those in kitchen areas.

What a story could be told by this classic homestead. I found it in 'Trentino Emigrazione', a magazine catering to tourists and Northern Italian emigrants. My first impression was that I had found proof that pole derricks were used in Europe. Under magnifying glass, it looks more like two men are carrying in loads by fork. What the two upright poles were used for is unclear.

Families and friends would gather in the evening in well-maintained loft and manger areas to share warmth generated by the animals. There they would while away time telling stories, sipping wine and meditating as winter cleansed the valleys for a new spring. In Val di Non, a high country valley in northern Italy, such gatherings were called filos.

Today many of these rustic quarters have been converted to modern bedrooms and store fronts but if you listen carefully, history will whisper its wondered past as you stroll by.

Two views of barn hay mow methodology: Fedele Asson, my grandfather, built this rock barn around 1929 near Rupert, Idaho. It had a 'bird beak' projection to house the unloader. A simpler rail extension with the unloader in place is shown on a barn discovered in the Wasatch Mountains near Huntsville, Utah.

ODE TO BARNS PAST

Conceived in need and built with pride,
by careful loving hands.
Mystique with styles of purpose blend,
from your ancestral lands.

Your mow and stall made food for all,
with plenty left to sell.
You've sheltered countless herds and flocks,
and served your masters well.

Technology or laziness,
It's hard to trace the blame.
For our neglect of your distress,
we all share common shame.

As future craftsmen reconstruct
your every joint and truss,
May history be as kind to you
as you have been to us.

A death no reason justifies,
a tragic way to end.
So as we say our last good-byes,
we must add, "Thank You, Friend".

Richard Talada – 1990
Barton, New York

Unexpected rewards sprang from this endeavor. Pleasantries occurred spontaneously. In my travels, I met accomplished people, learned of their doings and saw our country from a rare perspective. Touring western America in search of material, my eyes swept the countryside for barns, wagons, rakes, mowers and stackers. I also watched for cafes with pickups in the parking lot. Restaurant staff and guests within earshot were frequent sources of invaluable clues. It is amazing what one can learn through random dialogue. Suggestions flowed each time I stepped out of my normally reserved mode to talk with strangers, especially in off-the-beaten path locales. I was grateful for the data and ever inspired by the ambiance.

Tangy Bair served me coffee at The Frontier Pie Shop in St. Anthony, Idaho, one morning and provided directions to a back road to Ririe, all the while wondering why. "There's nothing there. Derricks? What's a derrick? Am I supposed to know? Ooooh! Now I remember something. Grandpa told me about my uncle hooking up a wash tub to the lift cable to make a crazy swing. One time the rope broke and he sailed into a hog pen." Wow! She gave me her grandpa's phone number. A year later he explained subtle distinctions between single and double 'A' frame derricks.

Scott Bridges, owner of the Lone Pine, directed me to an excellent photographic shoot of hard to find equipment. He even divulged the location of some hidden Indian rock paintings carefully preserved on the owner's ranch.

Back Country Dining

If ever nearby, check out these diners. Some are a little off the beaten path but well worth the time and I know you will enjoy the adventure.

Pastime Cafe	*US 12*	*Walla Walla, Washington*
Dirty Annie's	*US 14*	*Shell, Wyoming*
Yellowstone Drug	*US 20*	*Shoshoni, Wyoming*
Pine Inn	*US 20*	*Burns, Oregon*
The Wort	*US 26*	*Jackson Hole, Wyoming*
Attitude Chophouse	*US 30*	*Laramie, Wyoming*
The Hut	*US 30*	*Pendleton, Oregon*
The Other End	*US 40*	*Heber, Utah*
Red Barn	*US 50*	*Montrose, Colorado*
Shady Nook	*US 93*	*Salmon, Idaho*
The Loading Chute	*US 93*	*Carey, Idaho*
Chevron Deli	*US 95*	*Paradise Valley, Nevada*
Eidelwiess	*Cal 49*	*SutterCreek, California*
Lone Pine	*Id 28*	*Near Leadore, Idaho*
Kasino Club	*Id 21*	*Stanley, Idaho*
Horse Prairie Hilton	*L & C trail*	*Grant, Montana*
The Homestead	*Or 31*	*Paisley, Oregon*
The Timber Mine	*Ut 39*	*Ogden, Utah*
M & T's Bar & Grill	*Wa 14*	*Roosevelt, Washington*

Bon Appetito

A niece's wedding in Las Vegas provided another driving opportunity – through California's vast alfalfa growing center. California was a leading producer of alfalfa in the 1920s and still ranks in the top 5. Strangely, none of my California tours produced a single sighting of any old-fashion hay stacker. I found relics in every other state. There were horse mowers, rakes and loaders scattered about – and these usually presaged stackers – but none was found. Letters to colleges and industry didn't solve the mystery.

Visiting with wedding guests produced other information, however.

Verne Johnson and Neil Schmitt leafed through an early draft of this book, wondered at my literary sanity and fell into telling earthy tales of days when kids really worked. Machinery used in their boyhood days and high prairie antics were recalled in ribald detail. Such as one about a hard scrabble farmer out of Spokane who used a ramshackle stacker that nearly killed his crew. It was necessary to manhandle jags away from the mast of the boomless derrick, while the load was being pulled aloft. If they pulled too hard, the whole assembly tipped over endangering all in 60 feet. When the old coot stopped laughing at the ruckus, his crew had to re-erect the massive tree.

Scouting in northern Nevada's Paradise Valley, my wife and I stopped to examine a wasted wagon decaying in sagebrush at the end of a field. A farmer stopped to check us out.

Joe and Nancy Sicking proudly toured us through their turn of the century 3-story home and antique barn filled with treasures. Later Joe directed us to equipment sightings hidden to all but native eyes.

Turns out he is a history buff himself and invited us to see his ancient wooden barn complete with working trolley and carrier. The barn also housed a well preserved, belt driven seed cleaner, circa 1903. Incidently, on the wagon, were the remains of a swing-around stacker, a model supposedly not to be found in this locale. A wooden threshing machine stood nearby with hopper agape begging for a museum to rescue it.

In another exceptional coincidence, the mayor of Winnemucca was seated next plate to me at a family table in *The Martin,* one wintery March evening. Mr. Vesco and his wife interpreted the superb Basque menu while hearing of my interests. Shortly the wonders of Humboldt County, Paradise Valley and Nevada ranching flowed like sangria as they laid out a personal research itinerary with travel directions.

These were but a few lucky side lights – unplanned encounters that played evocative roles in the development of this book.

Separate barns with large haymows were obvious responses to growing demand. As crop volume grew, new methods were invented to put away the harvest more comfortably. Lifting equipment used in outside stacking – Jackson forks, grapples and nets – worked interchangeably indoors. The principal difference was trackage.

An excellent sketch, taken from the now familiar F.E. Myers and Bro. catalog, clearly sets out the inside components of a barn stacking operation. The scene opens with the team having pulled a loaded wagon inside the barn and rehitched to the carrier lift rope. Sling chains beneath the upper half of a two-tier load have been snap hooked to parallel sling pulleys. As the lift rope curls upward through the trolley carrier, the load will reach desired

height where the inside man will call halt, position the load with his fork and yell *'Trip It'*. Inside unloading is identical to outside 'bird beak' work except that loads from the outside, pass through a large drop down entry door on the way to the mow.

The un-loader carriage hung from tracks made of steel or 6" x 10" wooden beams that ran the length of the barn. Stops kept the carriage from running off the ends. A special catch or

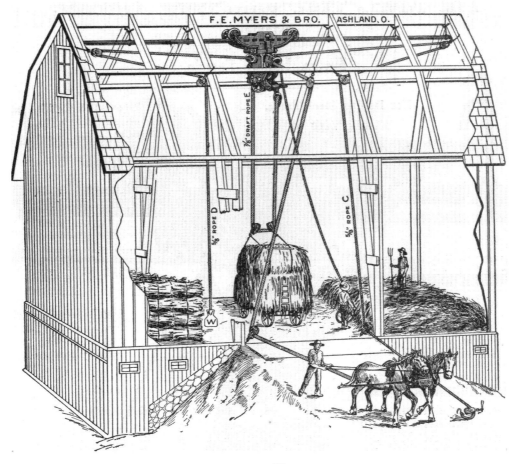

The Clover Leaf Right Angle Sling Elevator.

PATENTED.
FOR DOUBLE STEEL TRACK.
The Machine That Deposits the Load in the Mow in the Right Position Without the Use of Right Angle Pulleys.

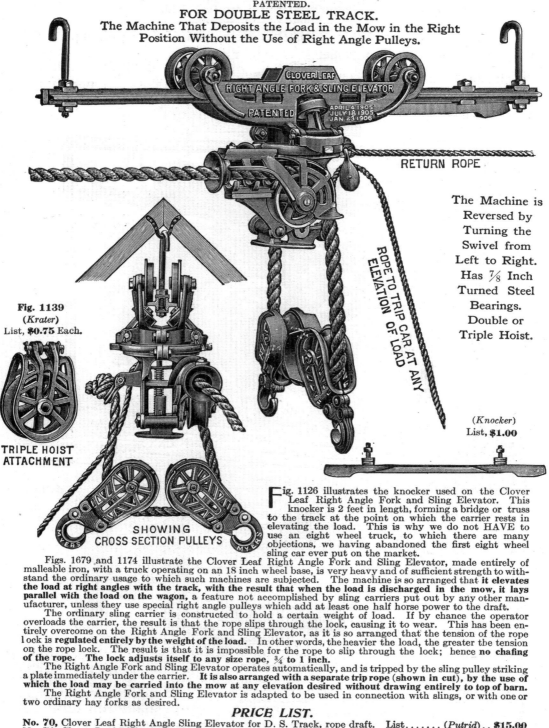

RETURN ROPE

ROPE TO TRIP CAR AT ANY ELEVATION OF LOAD

The Machine is Reversed by Turning the Swivel from Left to Right. Has ⅞ Inch Turned Steel Bearings. Double or Triple Hoist.

Fig. 1139
(Krater)
List, $0.75 Each.

TRIPLE HOIST ATTACHMENT

SHOWING CROSS SECTION PULLEYS

(Knocker)
List, $1.00

Fig. 1126 illustrates the knocker used on the Clover Leaf Right Angle Fork and Sling Elevator. This knocker is 2 feet in length, forming a bridge or truss to the track at the point on which the carrier rests in elevating the load. This is why we do not HAVE to use an eight wheel truck, to which there are many objections, we having abandoned the first eight wheel sling car ever put on the market.

Figs. 1679 and 1174 illustrate the Clover Leaf Right Angle Fork and Sling Elevator, made entirely of malleable iron, with a truck operating on an 18 inch wheel base, is very heavy and of sufficient strength to withstand the ordinary usage to which such machines are subjected. The machine is so arranged that **it elevates the load at right angles with the track, with the result that when the load is discharged in the mow, it lays parallel with the load on the wagon,** a feature not accomplished by sling carriers put out by any other manufacturer, unless they use special right angle pulleys which add at least one half horse power to the draft.

The ordinary sling carrier is constructed to hold a certain weight of load. If by chance the operator overloads the carrier, the result is that the rope slips through the lock, causing it to wear. This has been entirely overcome on the Right Angle Fork and Sling Elevator, as it is so arranged that the tension of the rope lock is **regulated entirely by the weight of the load.** In other words, the heavier the load, the greater the tension on the rope lock. The result is that it is impossible for the rope to slip through the lock; hence have **no chafing of the rope. The lock adjusts itself to any size rope,** ¾ to 1 inch.

The Right Angle Fork and Sling Elevator operates automatically, and is tripped by the sling pulley striking a plate immediately under the carrier. **It is also arranged with a separate trip rope (shown in cut), by the use of which the load may be carried into the mow at any elevation desired without drawing entirely to top of barn.**

The Right Angle Fork and Sling Elevator is adapted to be used in connection with slings, or with one or two ordinary hay forks as desired.

PRICE LIST.

No. 70, Clover Leaf Right Angle Sling Elevator for D. S. Track, rope draft. List....... *(Putrid)*.. $15.00
No. 71, Clover Leaf Right Angle Sling Elevator for D. S. Track, rope draft, triple hoist.
List .... *(Pvtt)*.. 15.75

A close up of Myers' Clover Leaf Right Angle Sling un-loader carriage gives much interesting detail as to how the main part of the carrier system works.

switching mechanism controlled and changed the upward direction of the load to send it laterally down the track when the elevating ropes touched the top. A smaller gauge rope pulled along by the carriage activated the tripper. After releasing the load, the wagon man pulled the un-loader back to the end of the track and down again for the next hitch.

Another Way

The photo above right shows another way of getting hay into a barn. A series of large doors were built just below the rafters of this simple two-story shed barn. The rest was automated. Think about it! Using a push rake to barn stack a load of hay. Sticking the tines directly through an open doorway, instead of lifting hay with a carrier, efficiently delivered a meal to the loft deck. Now, hay was gravity-ready for tossing into downstairs mangers affronting milking and feeding stanchions. It meant using smaller equipment and would not do for large dairies but sure handled the chore for a farmer with a modest herd and short on helping hands.

Russell S. Else left drought stricken Nebraska bringing this buck rake with him and thus introduced a new idea to his neighbors in Douglas County, Wisconsin. Another of the historical contributions from our nation's Library of Congress.

Sometimes after the hay was stacked neatly in the barn, folks wanted to know how much they had 'put up'.

The US Dept. of Agriculture developed the formula and John Deere Tractor Co. published it in a *Farmer's Pocket Ledger.* Idaho Implement

Not all is cut and stacked

Pasture crops are generally high in protein so long as they are kept growing and prevented from heading out. Another forage method is to cut and feed a crop green and fresh, i.e. just enough for the day's meal. This form, called soilage, (not to be confused with silage) is seldom used in today's more pressured world.

distributed the booklet in 1958 to Southern Idaho farmers. Here is how it worked:

To Find The Number Of Tons Of Hay In A Mow:

> *Multiply the length by the width by the height (all in feet) and divide by 400 to 500, depending on the kind of hay and how long it has been in the mow.*

In closing tribute to barns of old, recall what must be the fondest memory of hay mow days: the ecstasy of the daring, idle time jump – actual or imaginary – from the dusty rafters into deep piled mounds of fresh cut hay. Fantasy or fact, those were good olde days.

Then again: quietly reflecting in the soft, warm shadows of a retreating sun about another day of productive splendor was hard to beat. The gratifications of a day spent in farm toil were only exceeded by the anticipation of tomorrow.

Then again:

Farming looks mighty easy when your plow is a pencil, and you're a thousand miles from the cornfield.
Dwight D. Eisenhower (1890–1969), U.S.General, Republican President. Speech, 25 Sept. 1956, Peoria.

CHAPTER 7

Colossal Stackers

The giants in the odyssey of bringing in the hay – the colossal devices designed and built to stack hay in the open air – were the inspiration for this book. As a youngster, I was lucky enough to be around and take part as a stamper and wagon driver and later on as a full-fledged time farming scenes specializing in hay derricks. Many were shot on camping vacations with the family as we motored through Idaho, Wyoming and Utah. When they loomed in the distance, I braked to a stop, let the kids roam a bit and walked out with my Olympus for another treasure. There was no intent on writing then – just a fond interest in a curious

hay pitcher. Years passed. Nostalgia struck. I got interested in the arcane sport of derrick spotting and began collecting pictures of old-subject. How or when the idea arose to add a few observations is vague. Perhaps it developed while looking at old pictures and thinking up captions for a collage of the better shots – something to mat and hang on the wall in the den. More likely the urge to comment was tied to the timeless parental compulsion to bestow on one's kids an appreciation for the good old days. From there it just grew – a few remarks on nutrition, the intricacies of scything, how tedders changed agriculture – well, you get my drift.

From Sage Brush to Harvest by Irrigation.

Reclamation Act – Newlands Act of 1902

The community where I grew up was formed largely by The US Bureau of Reclamation. Irrigation water began flowing in 1909, thanks to a system of canals dug across the fertile but arid southern Idaho valley. Minidoka was one of 26 similar projects funded by the federal government at the turn of the century.

The Newlands Act of 1902, named for its author Francis Griffith Newlands, Democratic Representative from Nevada, created the

reclamation service which later became the United States Bureau of Reclamation. The bill authorized the sale of Federal lands with proceeds going to build power generating and water storage irrigation systems in sixteen western states. A peculiar feature of the act was to establish farms for the relief of urban congestion.

As an aside, Rupert High School, in the county seat of Minidoka, was the first school in the United States to be fully powered by electricity. I graduated with the Class of '55 – the last of the Pirates from that grand-old school.

Canals, dug with fresno scrapers pulled by horse and mule, channeled millions of gallons of gravity-flowing water to miles and miles of flat, sagebrush covered prairies. Word was released of irrigable land about to become available for easy taking. Land rushes drew

The Umatilla Oregon Project 1907
<---Cold Springs Reservoir under construction

Another of the 26 projects formed during the land grant era – the Oregon California Klamath Project – was undergoing a severe conflict over water supply and species preservation as this was being written. The clash idled over 1,000 farms. Ironically, John Wesley Powell, an early day environmental activist, was a sponsor of the Newlands Act and strongly supported these western reclamation projects.

Historic pictures on this page are courtesy of:

Photographs of the American West 1861-1912

National Archives and Records Administration Washington, DC 20408

hopeful families from everywhere. In a few short years, potato, sugar beet and alfalfa fields displaced thousands of acres of sagebrush.

Mile-square, 640 acre plats, laid out on a true North-South axis, budded with activity. Spud cellars and hay derricks were in demand for storing and stacking first time harvests. Almost every 80-acre farm sported one of each. The

half-underground cellars and tall standing derrick hulks became so commonplace as to seldom draw a second glance from farmer, banker or visitor.

A-frame derricks, such as the beauty in the photograph above, near Montrose, Colorado – the Uncompahgre Valley Irrigation Project, some 500 miles east of Minidoka – closely resembled those rising in Minidoka. Sixty foot long pine booms, hanging on heavy chains attached to the inside apex of the namesake "A", were an arresting characteristic. Solid. Practical. Unequalled. I was jaded.

Our next-door neighbors, the Walters, built stacks with a different model. The boom was mounted to pivot on top of a sturdy, vertical mast. These claimed the advantage of a wider service arc, since the boom could swing without bumping into the supporting triangular framework of the "A". In spite of close contact – I often worked for the Schenks, Ketterlings and Bonadimans, farmers who used pivot derricks – some childish partiality blocked the pivot's very existence from my memory. A-frames – that was it. And, since I had no then-current knowledge of the wider assortment of other stackers, I grew up thinking, as many others did, that derricks were the only kinds of hay-stacking devices.

More Than Derricks

But there were others. So many others! Even a look back at the obvious diversity in my own photo collection failed to alter my irrational predisposition. Until commencing serious research on this project, my narrow perspective held firm. What an awakening! A wider mix of specialty farming equipment is

hard to imagine. Heavy boom derricks were a major kind of hay stacker but far from the solitary medium. In this chapter, we'll explore A-frame and pivot derricks *plus fifteen other ways* to put up hay.

Here, at a glance, are a few of the many stackers used by farmers and ranchers throughout the West. Why there were so many and how they came to be is our story.

There were overshots, swingers, combinations, beaverslides, cable stackers, back flips, tripods, rope lines, pole lifts, wilsons, and Mormons.

Maybe even others, lost to early farming's recorded history.

Montana

Somewhere in Ohio

Kelowna, British Columbia

Nyssa, Oregon

Somewhere in Kansas

The Multifaceted Experiment of Putting Up Hay

Photos on this page and many others in this chapter are courtesy of the United States Library of Congress, Prints and Photographs Division. Several series, organized under the caption 'American Memory', present a fantastic review of agricultural history.
Find them at http://memory.loc.gov.

America from the Great Depression to World War II
Prairie Settlement: Nebraska Photographs and Family Letters 1862-1912.
Buckaroos in Paradise: Ranching Culture in Northern Nevada 1945-1982.

Introduction To Hay Stackers
Seventeen Ways To Put Up Hay

Description	Guy Wires	Wheels Skids	Dismantle	Capacity
Slide Stackers				
Sb Beaverslide		Y		Medium
So Plunger overshot		Y		Medium
Tooth Stackers				
Tc Combination rake/stacker		Y		Medium
To Back flip overshot		Y		Medium
Ts Swing-around		Y		Medium
Boom Derricks				
B1 Leaning mast	Y		Y	Light
B2 Jointed or chained pole derrick	Y		Y	Light
B3 Wilson derrick	Y		Y	Medium
B4 Turning mast with braced boom		Y		Medium
B5 Top mount pivot boom		Y		Heavy
B6 Single and double A-Frame		Y		Heavy
B7 Portable boom stacker	Y		Y	Medium
Rope, Cable and Trolleys				
Rt Tripod			Y	Light
Rc Suspended cable	Y		Y	Medium
Rr Rope pull			Y	Light
Rb Trolley carrier in barn				Light
Rf Common spike pitchfork				Light

Slide stackers came in two angled forms. One pushed hay up 45 degree inclines with long poled plungers. The other had pulleys at the top and pulled cradled loads up with cables.

Tooth stackers lifted loads on cradles made of long poles or teeth. The backflip tossed hay up and over itself. Swingers lifted loads along the stack side and then swung over the top. Wheeled combos could pick up hay in the field and set it on the stack with a charging run.

Boom derricks were constructed with heavy log boom poles mounted on tall frameworks. They could lift, swing and place gigantic loads at almost any point on a stack.

Rudimentary devices ranged from simple labor intensive pitch forks to cable and rope systems. They were precursors to the more mechanically efficient advancements.

Why So Many?

Seventeen ways to put up hay: Isn't that a bit much? The introductory outline on the facing page starts a study of why.

Greybull, Wyo.

Self descriptive model names, sorted by class, delineate the syllabus. In this section, serious students will learn fundamentals of progressive hay stacking together with the underlying reasons for selecting one stacker over another. Upon completion, scholars will embrace heretofore unrevealed particulars of key concern to pioneering users. The mystery of why so many came to be will be solved. We commence with a survey of the rudiments of stacking hay in the great outdoors.

Productive tonnage was of prime importance when considering which stacker to use. Small meandering fields were usually harvested with lighter, more mobile equipment than was practical for larger or more distant tonnages. Once a stack was topped off to a practical height, small, flexible stackers were dragged to new settings several hundred yards away. In heavy production cases, loads were usually brought in from far flung fields to a central stacking area on wagons or slips. Heavier stackers, capable of lifting large loads, but inherently less portable, were parked at the central stack yards. Moving awkward, heavy, wobbly apparatus any more than a few yards was a difficult task fated for damage and expensive repair when performed carelessly or excessively.

The kind of hay grown, peculiar as it may seem, was another big factor in stacker choice. The reason makes sense. Thick, coarse hay, like alfalfa, handled differently than fine, slippery meadow grass. Jackson forks, nets and sling chains worked well with alfalfa. The interlacing character of alfalfa stems held the shocks together so large loads could be gripped and lifted with each set. Pressing long, wide spaced, spring steel Jackson fork tines or slim barbed grapples into a load of high quality grass hay would have drawn more than snickers from a crusty, high meadow teamster. Beaverslides and long tooth swingers were designed to handle these silky cargoes.

Available resources – cash, trees and access to blacksmiths – also played decisive roles. The simplest stackers required some store bought parts – collets to affix booms, cable and pulleys to carry the loads, and manila hemp. Trade offs were involved. Guy wired

devices were easier to construct and less expensive but had to be dismantled for movement. Boom stackers were efficient for handling large loads, yet they were more expensive and took strong teams of horses to move even short distances. Combination push rake/stackers moved easily and could collect and transport hay directly from field to stack but had limited capacity. *Decisions. Decisions. Decisions.*

To make matters more interesting, this was not the only dilemma farmers faced. After grasping the intricacies of their circumstances, there was another question: How to go about getting a stacker? They would find it largely a do-it-yourself kind of thing. Some historical perspective will be helpful here.

US Department of Agriculture publications document that most hay stacking equipment had been developed and placed into service before 1900. H. B. McClure and L. A. Reynoldson compiled two bulletins while under assignment to aggregate and describe the state of mechanized stacking. Their efforts would produce the predominant source of information on hay stacking technology at the opening of the twentieth century.

Mr. McClure released Farmers' Bulletin 1009 in January 1919. The diminutive treatise with a lengthy title, *Hay Stackers, How They May Be Used in the East and South to Save Labor,* was the first to market. His ardent promotion of modern stacking was fervently held, as is evidenced in this lament about those reluctant to experiment with new ideas. He chided:

"Many Eastern farmers are cautious about spending money on something entirely untried. A few have tried stackers but have abandoned them, owing partly to inexperience and partly to difficulty in getting their men to give them

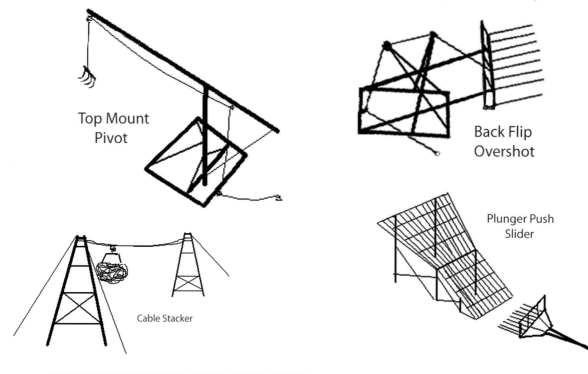

Top Mount Pivot

Back Flip Overshot

Plunger Push Slider

Cable Stacker

Genesis, logistics, artistic cunning and workmanship — all in the mind and hands of an enterprising stockman making improvements during the off-season.

a fair trial. Some of the stackers purchased were cheaply constructed and of a type not suited to conditions. Other farmers have not attempted to use stackers because they have been in the habit of making small stacks containing from 1 ½ to 3 tons, and think that hay will not keep in a large stack such as is made with a stacker – an objection not based on facts."

Above, a North Dakota backflip
Below, the crew posing with a swinger

His urging failed. Stackers could have been used in the East and South but they weren't. The Midwest participated using mostly slider and toothed models. But real acceptance of stackers was almost universally reserved to the West. One source placed the range of use from the ninetieth meridian (Iowa) to the irrigated parts of the mountain West.

Black and white photographs in McClure's bulletin hinted at key features of favorite models. The significant omissions were

beaverslides and some heavier boom stacker models that came into existence later in response to special needs and larger hay stacking volume. Brief usage descriptions complemented the photos and gave the reader an idea of what a completed machine should look like and how it worked. Details were skimpy. No working plans were included. Aspiring builders needed extensive, practical shop skill working from this slim resource.

McClure divided stackers into two general types: one utilizing long teeth that included overshots, combinations and swingers, and a second which he classed broadly as *'homemade stacking outfits'*. This was a catch

all for stackers that *"handled hay by means of the harpoon fork, grapple fork or sling."*

Reynoldson, an economist for the Bureau of Agriculture, opened his update, *Hay Stackers and Their Use,* stating that *"Hay stackers may be divided into four general types."* He retained McClure's long tooth category, added a slide stacker group, renamed McClure's *'homemade outfits'* to *'cable and derrick outfits'* and advocated a fourth type of limited value that used ropes to elevate hay. The revision added little text but did include more photos suitable for vague reverse engineering. Both issues were ripe with labor efficiency statistics comparing one model to another.

71

W. R. Humphries and R. B. Gray, cited earlier, recounted their USDA colleagues' earlier works, adding emphasis but not disclosing where to place an order. They did confirm that no evidence existed to clearly establish when or how stackers came into use. Tidbits scattered in agricultural textbooks, annual government reports and nostalgia magazines proved equally unrevealing.

Note that with one exception, all inventions were concocted by Midwestern minds. The latest recording was in 1869. No patents appeared in this series after that date. Eleven years of development without significant sales must have taken a toll. The stated purpose column didn't clearly reveal what style was patented but the preponderance of photos and advertising posters found in Midwestern

Recorded Patents Issued For Hay Stackers
From the 1949 Compilation of W. R. Humphries and R.B. Gray

Patent Date	Inventor	Address	Stated Purpose
Dec. 28, 1858	C. W. Glover	Farm Ridge, Illinois	Stack agriculture products
Dec. 28, 1858	J. Van Doren	Farm Ridge, Illinois	Stack agriculture products
Feb. 25, 1862	W. M. Mason	Polo, Illinois	Hay stacking machine
Oct. 18, 1864	F. Wickes	Kansas, Illinois	Hay elevator
Jul. 11, 1865	A. W. Tooker	Harvard, Illinois	Hay elevator and stacker
Apr. 24, 1866	C. Rundell	Chicago, Illinois	Hay and straw stacker
Jul. 10, 1866	C. Rundell	Chicago, Illinois	Hay stacker
Aug. 8, 1866	W. Louden	Fairfield, Iowa	Hay stacking device
Jan. 8, 1867	J. T. Breneman	Springfield, Ohio	Hay stacker
Jan. 29, 1867	C. H. Tryon	Greenwood, Illinois	Hay stacking apparatus
Apr. 23, 1867	S. J. Wallace	Keokuk, Iowa	Hay stacker
Oct. 29, 1867	Forshee/McCland	Unionville Ctr, Ohio	Hay stacker
Jan. 28, 1868	W.S. Nichols	Rutland, Vermont	Hay stacking equipment
Oct. 19, 1869	T. N. Bunnell	Reynolds, Indiana	Hay stacker
Nov. 30, 1869	J. R. Hammond	Sedalia, Missouri	Rake, loader, stacker

Industry was not the answer. Except for a weak effort in the mid–1800s, few commercial ventures were attempted. Humphries and Gray's *Partial History of Haying Equipment*, included a listing of patents. Brevity was its most interesting feature. It spoke volumes about the commercial derivation of stackers. The authors further dismissed the impact of the few businesses that did obtain a patent by saying *"it is unknown how many of these machines were ever manufactured commercially"*. In my travels, I only saw a few swingers that may have been 'store built'.

historical archives are of swinger and overshot varieties. Most likely, the majority was of the toothed stacker class.

Clearly people were growing tons of hay and using a multitude of stackers to harvest it. Materials published by government and private parties told a lot about the existing situation but offered little source information. Perhaps the answers as to why so many stacker types existed and how they came to be had been lost to history. I was about to give up. Then I found another book.

Turn of Century Manufacturers

John Deere Company, 1876 , Moline, Illinois
  Mowers, rakes, loaders, overshots, swingers
    and combo/rake stackers

Dain Mfg Co., Ottumwa, Iowa   Buckrakes

McCormick Deering, 1848, Virginia
International Harvester, successor, 1902
  Grain reapers, mowers, rakes, loaders, tedders

J.I. Case Company 1842, Racine, Wisconsin
  Steam engines, threshers, tractors, rakes

F. Wyatt Company, 1904, Salina, Kansas
  Jayhawk stacker, combo stacker, sweep rakes

Jenkins Hay Rake and Stacker Factory
  Chilicothe, Missouri, Jenkins swinger stacker

F. E. Myers & Bro., 1870, Ashland, Ohio
  Unloaders, pulleys, grapples, nets, trippers

Aultman, Miller & Co., 1864 , Akron, Ohio
  Buckeye brand grain reapers, mowers, binders

Folklore

Pioneers took to the Oregon Trail in 1841. Mormons left Illinois for Utah in 1847. Civil War battlefield hostility ended in 1865. The Golden Spike opened rail service in 1869. As people looked for greener pastures, they took what they could carry, leaving behind heavy immobile items like windmills, derricks and sheds. Fetching along heavy cargo was beyond contemplation. Getting there intact and figuring out what to do next was hard enough.

The study of cultural geography, and more particularly folklore, attempts to shed light on how things circulate from place to place.

Folklore is defined as the comparative study of folk knowledge and culture – the traditional beliefs, myths, tales, and practices of a people, transmitted orally. The focus of a cultural geographer's study is the regionalization or commonality assumed by distinct apparatus within a community and the linkages retained in an item found in a subsequently developed neighboring community. The patterns can suggest the reason for modifications and the route of diffusion of the artifact.

Whether the study is of banjos, barns, food or derricks, the idea is to see how change happened and how the item made its way around the world. The science of cultural geography/folklore could be applied to hay stackers according to two Utah scholars who shared a special interest in the dissemination of boom derricks.

Utah hay stacks with leaning mast stacker at left

Courtesy of Library of Congress

Vermont hay stacks likely put up with pitch forks

73

In a too short but absorbing article entitled *Hay Derricks of the Great Basin and Upper Snake River Valley,* Austin and James Fife issued what has to be the premier folklore report on the assortment of hay derricks found in Utah and Idaho. Their 1948 study was conducted along a thousand mile stretch of US Highways 89, 91 and 191 from Bunkerville, Nevada through Utah and Idaho, to Yellowstone National Park. These avid fishermen paused between fishing holes to count, categorize and regionalize over 1,500 hay stackers. Talk about getting into derrick spotting!

Six types of boom derricks were charted from roadside observations as the Fifes motored along. They considered their motorway study statistically very thorough, commenting: *"Nearly all the irrigated valleys between Bunkerville, Nevada, and Salt Lake City are narrow and are separated from each other by uncultivated areas that make observations from the highway highly reliable. More than 80 percent of all the derricks in the valleys touched by the highway were counted, except in the Utah and Salt Lake valleys and in the restricted area around St. George, Utah, which we were forced to pass through in darkness."*

Recall the broad classification of stackers presented at the outset of this chapter. As comprehensive as the Fife brothers' observational study was, it examined only one of the groups – *boom derricks*. The finer points of these trend setting machines are disclosed on page 76.

Austin and James Fife were faintly aware of non-derrick devices but disregarded them because so few were found. Under pressure,

in a subsequent update, they recognized the *"...intrusion from other areas of two stacker types into the region of our investigation: the 'beaverslide' and the 'overshot'. These types, seemingly of folk origin and development also, do not originate in Mormonia, but their efficiency seems to be such that they may in time replace the 'Mormon' family.* The reference to Mormonia seemed to me a coded clue. The incompletely developed passage intimated that credit may be due to migrating members of the Church of Jesus Christ of Latter-day Saints, at least for certain kinds of pole derricks.

Dean May, writing a newsletter on Latter-Day Saints history, made the connection. He proposed that the *"... juxtaposition of modern attitudes toward farming, skills gained in early industrial Britain, and the pressing need to increase production on Utah's hardscrabble farms ..."* led to derrick development. Gangly structures of significantly dissimilar appearance became known throughout Utah and nearby states as 'Mormon' stackers, using the church's common appellation.

This picture was included in May's article with the following subscript: *"This style of hay derrick (c. 1900 on Blue Creek Ranch, near Brigham City, Utah), introduced into the area by Danish converts, became widely known as*

the 'Mormon derrick'." Pole derricks such as the one this *'jolly crew'* was using were simple affairs. Heavy log chains snubbed the boom to the upright mast which was planted in the ground like a tree and held erect by guy wires not clearly visible in the picture. Sling chains bolstered the half-ton load. The winged nut on the boom had no structural significance.

A super-structured variation was photographed near Joseph, Utah, some 230 miles south of Brigham. Extensive triangular bracing strengthened the boom and increased capacity but since the angled supports were bolted to both the boom and the mast, the whole assembly had to turn inside its cross-braced frame. Despite the obvious differences, period writers referred to both as 'Mormon' derricks.

The common link between the two models was that both employed heavy timber and used pulleys to raise the load. It is believed that knowledge and experience gained working on industrial cranes in European ports or aboard ships played a role in the preliminary design of boom stackers. Transitional improvements led to the substitution of structural bracing for guy wires used on early models. These examples and other boom-type derricks are therefore considered related through the developmental process.

Once discovered, derrick proliferation was aided by newspaper articles such as this from the August 27, 1908 *Pioneer Record*:

A Derrick Which the Farmer Can Make for Himself

The derrick consists of two principal parts, a revolving boom pole or crane, which swings in a complete circle, and a framework for supporting the crane. The crane part is made with an upright mast 25 feet long and ten inches in diameter at the base, with an iron band and iron pivot pin at the base; pivot pin should be 1 ¼ by 12 inches or larger.

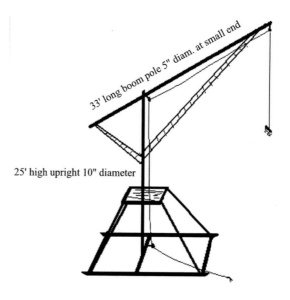

33' long boom pole 5" diam. at small end

25' high upright 10" diameter

The mast supports a boom pole 33 feet long, five inches in diameter at small, upper end. It should be bolted to top of mast, 11 feet from the butt. The top should be about 32 feet from the ground when in position. This boom pole should be supported by a pair of long poles for the top and a pair of shorter poles at the butt as braces. The butts of these braces should be bolted to the mast at about the center, as shown in diagram. A pulley should be hung at each end of the boom pole.

The base of the supporting framework should be of two poles about eight inches in diameter and 18 feet long, the underside of each end rounded up, sled-runner style. These poles should be 15 or 16 feet apart. Near each end

Derrick Types of the Great Basin and Upper Snake River Valley

A Survey by Austin and James Fife from Western Folklore Volume 7, 1948

Type	Derrick Symbol	Description of Derrick Type
1 B1	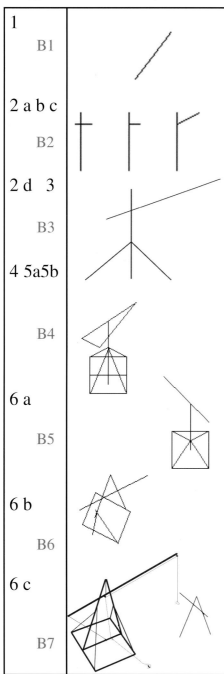	The simplest of all derricks was a solitary leaning mast thinly anchored in the ground and tilted to extend over the stack. It was supported by 3 or more guy wires and needed only two pulleys. The load was dragged up the haystack side.
2 a b c B2		Three styles of pole derricks differ only in how the boom was suspended over the stack. Guy wires were the main support for the upright mast though some models had light base bracing.
2 d 3 B3		Heavier base bracing differentiated enhanced pole derricks from models 2 a, b and c. Early boom poles were simply wrapped to the upright mast using heavy log chains. Models using more sophisticated joinery made from steel and fork shaped hardwood were known in some locales as Wilson pole derricks.
4 5a5b B4		Type 4 derricks adopted a significant design change. The base of the upright mast was moved away from the side of the stack and rigorously buttressed by log braces. Guy wires were eliminated but a longer and more sturdy boom was required to span longer reaches.
6 a B5		Boom bracing is the obvious difference in B4s. The maladroit subtlety of 'Figure 4' derricks was that the mast supporting the boom had to rotate in a socket at its base because the logs bracing the boom were solidly fixed to the mast. I suspect it was a squealer.
6 b B6		Top mounted booms, colloquially known as fish poles, provided wide arc pivoting. Specially fabricated metal collar and pin assemblies secured the boom, which rotated in a socket at the top of the mast. Greasing was necessary.
6 c B7		A heavy horizontal cross beam connected two triangular uprights on wide body, double A-frame configurations. The boom hung by chain from a center point and was practically unrestricted in lateral movement. Wide expanses and limited cross bracing made some vulnerable to twisting and damage under heavy swinging loads.
		The classic single 'A' frame was slightly more movement restrictive but proved the most stable and efficient of boom-type derricks. Lighter, one-sided, portable 'A' versions used guy wires to stabilize uprights. Numerous variations existed for all models.

The Fife brothers' six-type categorization scheme dealt only with boom derricks, one of four broad categories of hay stackers. It is believed that knowledge and experience gained working on cranes in European ports or aboard ships played a role in the preliminary design of boom stackers. Routine transitional improvements on advanced models led to the substitution of structural bracing for guy wires used on early day poles. Symbols and codes used by the Fifes were modified slightly for classifying information in a computer data base developed by the author to analyze the distribution of all kinds of hay stackers.

a cross piece about six inches in diameter should be bolted, also a heavy cross stick across the center 10 or 11 inches in diameter. The upright mast is pivoted to the center of this stick, forming the lower bearing.

The upper bearing should be about 11 feet higher, made of a platform, four feet square of plank two or three inches thick. This is supported by posts about five inches in diameter, the base of the posts bolted or toe-nailed to the sled runner, near each end of runner and about 14 feet apart. The tops of the posts support the upper platform of plank. The center of the platform is cut out to receive the revolving upright mast forming the upper bearing. It is better to protect the mast at this place by a thin steel plate of about 6 to 12 inches or even better to have two plates at bottom of the mast with holes for the pivot pin.

To give the supporting framework additional strength and stiffness, explains Hoard's Dairyman, it is usual to attach two long braces from the bottom of each of the four sides to the opposite upper corners. A rope of pure manila three-fourths or seven-eights inch size is large enough. If harpoon or other horse forks are used three pulleys and 80 feet of rope are sufficient; 115 feet and four pulleys, if slings are used.

Nothing new appeared until February, 1946, when Oregon State College issued a circular discussing a guy wired boom-type stacker. Arnold Ebert, a county agent out of Condon, Oregon, joined with Clyde Walker of the Agricultural Engineering Department to promote an easily dismantleable unit. They said it easily outperformed conventional stackers in narrow valleys where moving on narrow roads was a real problem. Tom Huntington, a foreman on the C.H. Burgess ranch in Wheeler County, invented this unique stacker with but a whiff of public notice.

Portable boom type stacker, Oregon State College Extension Service. Circular 480, February 1946

A patent was issued in 1910 for a truly different model – one that handled hay using ramps instead of booms. David Stephens and Herb Armitage invented and patented their specialty in 1910. They named it a Beaverhead County slide stacker, or Beaverslide, in honor of their south western Montana county. Terrain and the type of hay grown were key design factors.

77

High quality meadow grass is the hay crop of Montana. Narrow valleys bounded by mountains are dissected by trout filled streams carving numerous irregular sized fields. The combination of light textured grass growing in small disjointed patches called for an easily transportable stacker that could handle large volumes. Other derricks were tried but none sufficed until an ideal model was discovered. The Beaverslide became so popular that the region is now known as *'The Valley of a Thousand Haystacks'*.

An historical marker proclaims a proud legacy. It stands on US 12 west of Helena abreast a brand new log beaverslide on the John Beck ranch near Avon, Montana. The stacker is parked for the off-season inside a square shaped stacking cage. Compare it to the round shaped model below. Nearby ranchers, Charles and Rita Gravely, told us that they put up 1,300 tons in 62 stacks this year. They used a steel model built in 2001. "We're in the Dark Ages but people love to come and watch us. It works for us and it's a great way to keep our past alive."

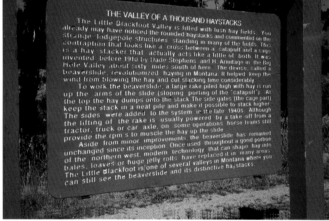

In Track of Stackers

As the widening makeup of stackers emerged, curiosity tugged harder. A more in-depth understanding of stacker distribution beckoned. Serious examination would require exploring many more valleys. But a lot of loads had been hauled before this idea occurred to me – or to anyone else, it seems. Passage of time caused the passing of many sightings. Sprinkler irrigation consolidated small fields eliminating previously mandatory but now wasteful ditches. Balers swept over haying landscapes tying ever larger and more compact

bundles. Stack yards were converted to more useful purposes, their raison d'être having vanished. Some disappeared through dispassionate riddance of no longer needed junk. Others simply collapsed in weathered demise.

Yet some remained. I had photographs which I had taken over the years. Historical museums held more. Could it be possible to rediscover the pattern existing at the height of stacker employment? I hoped so.

Given the luxury of highway improvement and by foregoing the *'fish while you study'* distraction, it was easy to cover a lot of ground. In time, I managed to drive most major routes and a lot of the back roads of the West. For those I missed, particularly those stretching deep into the Midwest, the Internet and gracious historical societies had answers.

Research was directed at the most promising hunting grounds. A graphic of America's Cattle Trails, sold by the Cowboys Then & Now Museum of Portland, Oregon helped show the way. Amazing how the old-timers laid path through water and forage. The Bureau of Reclamation map covering irrigation projects of the West suggested a similar approach. Charting stackers in juxtaposition to irrigable water hinted at why one type was used in the valley while another found preference on the drier side of the ridge. Everything counted.

The West is a big place though, with a lot of dry spots, basaltic rock flows and wide-open spaces not generally hospitable to growing hay. Remaining stackers are scattered widely. They are easy to spot when you sneak up on them but it takes a lot of time and patience. 'Stacker bingo' is not for the restless.

All stacker sightings made, whether by the Fifes, by myself or obtained from the Internet, were input to a computer database. Each bit of data obtained was saved for analysis. Data fields such as state, county, nearest city, highway number and milepost were entered. The type of hay grown, the farm or ranch name (if spotted on a gate post) and comments about a particular sighting were also recorded.

The Fife counts, made at the height of stacker usage, numbered far more than mine but were indispensable records. I had to be content with scattered onesy-twosy spottings. Would this present a statistical anomaly? Would charting my sightings next to the larger counts skew results? Could any observation accurately denote the predominant selection? Was it logical to assume that a particular sighting was the chosen model in a locale, purely by virtue of finding one still remaining in the valley?

I believe a positive answer can be defended. My sightings in Utah and Idaho made along

How did they build it?

This captivating photo, courtesy of Canada's Kelowna Museum, presents many questions. How did they build this stack? Was it a cable system between the trees? Did they take the cable down before the shot? Had they moved the derrick? Was it a hard-core 'pitchfork from the wagon to the stack' situation? Is the man on the stack showing off with a large shock? Or?

How many tons?

To find the number of tons of hay in a stack multiply the overthrow (the distance from the ground on one side over the top of the stack to the ground on the other side) by the length, then by the width (all in feet); multiply by 3; divide by 10 and then divide by 500 to 600, depending upon the length of time the hay has been in the stack.

$$\frac{\frac{(OV \; x \; L \; x \; W) \; x \; 3}{10}}{550}$$

The man on the right looks to be about 6 feet tall. Using him as a gauge I get 34 tons. What say you?

Starbuck – Touchet – Gardena

Two Washington farmers, 93-year-old Albert Accuntius and his cousin Ken Byrnes, delighted in long-lived careers involving alfalfa. These self-effacing farmers, mechanics, cowboys, soldiers, herdsmen, innovators and now historians conveyed a singular sense of pioneer resiliency to me one weekend in November 2002. On a nostalgic tour of their closely nestled communities, uniquely stirring accounts were unveiled in compelling recollection.

Ken Byrnes guided our morning tour through 'his' southern Washington valley proudly recounting scarcely known facts of a once world-renowned hay seed crop called Gardena. The common vex of water scarcity, combined with bogs of naturally proliferating alkali bees and a Los Angeles engineer named E.C. Burlingame, were background to Ken's story. Burlingame was called in to solve a perplexing canal washout problem. Riding horseback, he found a way to secure a 32-mile irrigation channel to the sandy-loam hillside. Walla Walla River water poured freely until Oregon won a lawsuit to halve the supply. Orchards planted in anticipation withered. Folks struggled for other livelihoods before stumbling onto an answer. Long, water-finding taproots allowed alfalfa to thrive on limited resources. More remarkably, a rare species of black bees needed for pollinating seed plants inhabited the area. The valley thrilled to overnight success. Byrnes pioneered the new seed market. As usual, there was more to the story.

A friend, who knew of my writing, mentioned derricks to her uncle one day. He told her he could tell me all about hay stacking, if I ever cared to call on him. Frances's uncle and Ken's cousin turned out to be vivacious Albert Accuntius. He was born on a hay ranch 50 miles northeast of Gardena near the booming railroad town of Starbuck. His father, another progressive of the time, was using pivot derricks on their 1,240 acre spread when the local John Deere dealer convinced him that a Jayhawker would serve better. It didn't. They settled on a cable outfit to build 100 ton loaves. This was a most unusual equipment transition – from pivot to jayhawker to cable – especially in a state where alfalfa ranked low both in acreage and production in 1920 – barely 3% of total USA production. I had to learn more.

In all my travel and research, I had never encountered a real life user of cable stackers. A postcard found in a book store proclaimed "Stacking Alfalfa, Washington" on its pictured face but I considered it just another photo in my collection. When Frances Accuntius introduced me to her Uncle Albert, who in turn shared the improbable details, I recalled the postcard. He recognized the cable stacker immediately as the kind they had used. Nothing indicates where the picture was taken but almost certainly it was on

Stacking Alfalfa, Washington.

the Accuntius Ranch in Starbuck. The passenger train is gone now. The rails were torn out for wartime steel. But in its heyday, the Washington Whistle Stop steamed alongside the ranch. The sheep dog chased it every morning as it passed. A promotion agent could have easily walked the ¼ mile to the ranch and shot the picture during a one-hour station break.

Behind the men's vivid stories, so relevant to this book were the real happenings. Neither had visited their places recently. Both sold out long ago and retired to Walla Walla. They see each other often and talk of old times but seldom to a new friend with a farming interest. The opportunity dug up long lost memories. In the overgrown 'bone yard' on Ken's old place, we found the first tool he welded with shaky beads – a paddle style post hole digger that worked better than any store bought auger; a forty foot wide, 4 seat dodder sprayer; a buck rake with dual offset wheels that cushioned travel over rills in corrugated fields and the remains of a Jenkins swinger the family bought in 1943 for $325.00 from the local John Deere dealer. "People were resourceful," Ken said. "Had to be. There were no manufacturers in the West that knew our concerns at the time. Inland Iron Works was around but they only did grain combines. If you needed something, you made it yourself. Oh my gosh, look here, the differential we turned upside down to reverse drive the buck rake and here …. Look at this …. And here …. for heaven's sake my whole life is scattered around this weed patch. For heaven's sake ….."

That afternoon we crashed in unannounced on the new owner of Albert's boyhood home. Mrs. Bishop reluctantly but graciously showed us in. When a 93 year old man announces on the porch steps that he was born in this place and wants to see it one more time, it is hard to say go away. The Bishops have lovingly restored the home and contributed much to the community but Starbuck is not as it was in Albert's memory. The school is boarded up. The hotel is gone. The restaurants and library have faded away. Even the dance hall has fallen down. It was such a lovely day – a grand tour with lively conversation. "But in the end it was like going to a funeral," Albert whispered ever so wistfully. At dinner that night he said, "Put that in your book."

During the day a young farmer drove up abruptly in his 4 x 4 to see who was invading his back road. Recognizing the old-timers he relaxed and carried on a light banter about rain and drought and wheat prices. "I am not sure what to pray for," he cynically worried. "Rain or drought, which is best? Government insurance almost beats the price of production."

It is a mark of our times and a fact of history. I know that the man who now owns the 'old Byrnes place' was bantering in jest. I know in quiet moments he appreciates the effort expended by his predecessors to bring water and life to his ranch. And I know that Ken and Albert wouldn't change much, even if they could. But the contrast sure made an impression on me one day in Touchet.

Myers Cable Unloader for Stacking in the Field.

Material Required for a Cable Outfit for a 50 Foot Stack:

One cable unloader.
One hay fork.
150 feet ½ or ⅝ inch galvanized cable.

Two single cable clamps, Fig. 431.
Two double cable clamps, Fig. 432.
Two ⅜ x 10 inch carriage bolts for fastening poles together.

130 feet ¾ inch manila rope.
Two pulleys.
Four poles 4 x 4 inches x 30 feet long.

routes paralleling the Fife Brothers' travels closely replicated their observations. Models spied in other states were frequently corroborated by evidence found in historical archives, museums and letters from knowing folks. Even popular literature validated.

Rick Steber, a famous author out of Prineville, Oregon, has written many books and articles about the early settlers of the West. In a 2002 release, *Buckaroo Heart,* he describes haying on the very real Steele Swamp Ranch in the Devil's Garden country of Southern Oregon. One passage states: *"Betty watched the buck rake dump hay at the bottom of a slide, a wooden ramp positioned at the foot of the stack. A net made of light chains was in position in front of the slide and hay was moved onto the net." "While he [the net pullback man] repositioned the net for the next load, the stackers pitched hay, building the stack, keeping the sides straight, rounding the top so the hay would lay flat and shed the rain and snow. Stacking took skill."* Mr. Steber may have been describing an incline plunger-push slider. This could be the one.

One of my trips took me through Devil's Garden. On the way, I stopped at an antique store in New Lone Pine Creek, Oregon, near the California border south of Lakeview. That is where I found the slider photo. It is reputedly from the files of a major local area ranch. The same antique store yielded another prize – a guy wired pole used on the nearby Heryford & Greer Ranch. Sliders and poles just down the road from each other. What were people thinking?

In another passage in Steber's book, an older buckaroo and drifter named Fred Westfall opted to leave Devil's Garden for greener pastures in Nevada. Fred tried a little jawbonin' to get the main character to move on with him. Said he wouldn't mind the company. Herman declined so Fred set out alone on his mount with a packhorse on a 10 day trip to Paradise Valley, Nevada. Paradise Valley is a quaint, remote oasis in northern Nevada's Humboldt County. It is strikingly familiar to me, however, thanks to a documentary from the Library of Congress called *Buckaroos in Paradise: Ranching Culture in Northern Nevada.* Leslie J. Stewart produced the audiovisual tape, that vividly recounts Nevada's loose hay stacking days.

I drove through Winnemucca and Paradise Valley sixty years after the time portrayed in the video to check things out for myself. The same trip took me within shouting distance of the Pitchfork Ranch where Steber's cowboys made their early life decisions. To stumble on this actual happening in casual, non-research reading was one of those special extras. Somehow I just have to believe that Fred was one of the hands in the documentary.

My Ford pickup and I covered 900 miles of Nevada, California and Oregon on that tour to locate 9 standing stackers – six sliders, one A-frame and two swingers. Archival data is critical to this kind of study. I was content to find nine 'live ones' but especially pleased to tie into the documenting passages about genuine Westerners settling real sections of Oregon and Nevada.

As you can see, my data sources were unrestricted. I have scouted over 20,000 miles in 15 states, visited dozens of stores, libraries, museums and historical societies and talked with tradesmen in the field. Information was rigorously collected and documented. I make no claim to scientific purity but believe the information is historically representative. Up to 300 line items in my data base can be sorted

Marvin Meek, Descendant of Joe Meek

William Meek, father of Marvin and distant relative of the famous mountain man/trapper Joe Meek, moved to Oregon around 1925. Marvin thinks he moved from Ashton or Rexburg – , "one of those Mormon towns up there."

In Idaho, William worked with pivot derricks and memorized enough construction details to become an expert. To make ends meet, he contract built 'fish poles' for the folks in the Cloverdale community between Bend and Sisters, Oregon. Marvin recalls helping haul jack pine logs from the Three Creek Lake area on the east flank of South Sister Mountain. A great folklore example!

Mormon derricks are not typical for this part of Oregon. When I spotted this failing carcass on a casual, property searching tour with my grandson, I had to inquire. Marvin continued to clean battery terminals with a hand file as he told us his story. Said it was a long time since he was born in 1919 on Oregon's famous Hay Creek Sheep Ranch up by Redmond but he is still managing to hang on and do his own repair. "Careful now. Don't nudge the old gal too much while taking her picture. She'll likely fall on ya."

by state or county, by class or type, or lined up along a particular highway. One version is tucked away in the appendix for those prone to elementary inquiry.

A color-coded map on the inside cover of this book and a topographic display on page 86 capture the distribution of stackers in compelling force. Spend a few moments assessing the pattern. Visualize terrain, soil fertility, altitude and the growing season. Think of water rights, available materials, shop skills and services at hand. Recall that most farming valleys are small chunks of tillable acreage nestled between mountains, basaltic lava flows, arid deserts, lakes and rivers. Bear in mind how terrain can change radically in 5 or 10 miles.

You are focused to understand why.

A Quick Review Before the Final Exam

➢ Hay stackers were found over a 1,700-mile front between the 90th and 122nd meridian – roughly from the right edge of Iowa to Oregon's high desert.

➢ The preponderance of use centered on the Great Basin territory formed by Idaho, Utah and Nevada plus a couple hundred miles along bordering states.

➢ Virtually all models were home built. Most of the devices thrived for barely 100 years; only one/twenty-fifth the known lifetime of alfalfa itself.

➢ Alfalfa districts, dominating wide, water-rich valleys, produced heavy tonnages requiring sturdy equipment. Of the four broad hay stacker groups, boom stackers served best. Pivots and A-frames were the giants in the field.

➢ Swingers and sliders were favored in grass or mixed hay producing areas.

➢ Cable stackers efficiently fashioned 50-foot long ricks. Tripods were convenient for small, low labor outfits. Combination rake/stackers were versatile. They could collect a load in the field, approach the stack from any direction and set the contents neatly 20 feet above ground level.

➢ Hayracks and slips were more likely used with boom stacker operations delivering 3-ton loads from long distances. Buck rakes galloped loads from nearby fields to the extended waiting teeth of beaverslides and overshot lifters.

Knowledge of what worked passed from one farm to another in casual 'boot-on-the-fence-rail' conversation. Little qualified as reliable documentation. There were no blueprints or construction guides similar to *'How To Build A Deck'* pamphlets found at local building

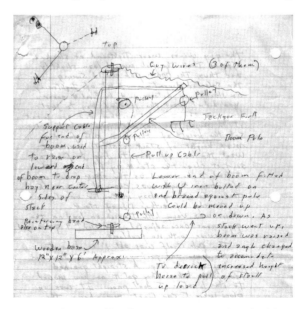

Mr. Rod Donnelly of Spray, Oregon, sketched his B2 and sent it along with a great letter on early hay stacking days. The letter is in the appendix.

84

Rankin was a name associated with commercially made backflip stackers. No further details could be found about the company. This model was being tested in Nebraska in the late 1880s.

stores these days. People considered their neighbor's apparatus, conjured up some personal improvements and scratched out a design in the stack yard dirt. A team of strong horses was hitched to the longest wagon and clucked to the woods where a sharp ax laid claim to the primary building material.

Practicality was the enduring guide. Our ancestors led remarkable lives under trying circumstances. Particularly remarkable was how they fashioned success through imagination. Ingenuity and economics were the starting points of every stacker design.

Mormons drew on their past to adapt crane technology for hoisting alfalfa instead of cargo. Early models were crude. Improvement came with experience.

Montanans faced a different problem and handled it in a singular, workable way.

Plainsmen tested industrial age ideas, and found value. But, facing practical economic circumstances, they concocted sensible ways to turn out home made models.

Rural America was scattered over a huge landscape in the days of derricks. Communication was not as we now enjoy.

85

Hardy stock considered glitches as imagination testing opportunities to be resolved locally. In a nut shell, the answer to why so many stackers came into use was a simple composite. People facing different problems in different locations dreamed up special ways to fit their special needs.

Choosing a hay stacker had to do with what one had to do and what one had to do it with.

Necessity, as always, was the Mother of Invention.

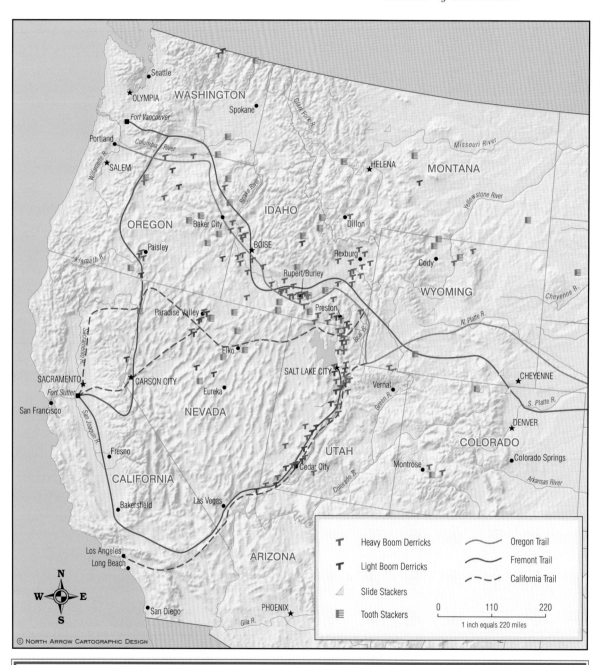

As I drove and learned, I became able to sense whether a stacker might be found in an area and what kind it should be. Folklore, economics, imagination – the factors that played key roles in derrick selection – came into better focus. Displaying the data topographically gives a feeling of how and why it happened.

Keeping Time

I remember a small farm known as the Grisenti place. It was in an ethnic settlement area referred to locally as the Emerson District. The immigrant farmer living there had a way with rocks. His barns and sheds were all made of rock. Even his irrigation ditches were lined with lava. When we were kids, Mom forced us to stop by after church to visit with these people whom I then considered stuffy old relatives. We children said our hellos and escaped quickly to play hide and seek around the rocks. Now I wish I had paid more attention. Maybe even taken a picture.

Dallas Carotta recalled driving a team to Hegler Canyon for derrick logs. One of his horses had a heart attack and died on the trip. They dug a grave on the mountainside. A challenging trip for timber and not a single photo. Uncle Ernie bragged over a glass of wine that he made two trips a day – a long day to be sure – to the same sawmill. Impossible for any team, considering the 80-mile round-trip pull, but always a thrilling story. Wouldn't it have been nice to watch him in

action? Now Gary Schorzman, as museum director for the same south Idaho county, guides history buff volunteers in search of new material to commemorate the spirit of unsung settlers. Why didn't I jot down more reflections as I rode bareback on old Rusty?

This was a question I asked myself often through this venture. In contemplation now, however, the pleas for more may be hollow. A discrete absence of data is fundamental to nostalgia. Excess image imperils the quest.

Treasured photos, diary notes, artifacts and 'things' are truly indispensable to keeping a connection with the past. *But there is more. So much more!*

People. People doing what people do.

Without deeds and doings, the 'things' left behind do little to mark a lifetime.

Deep felt glimpses – the longing expressions of folks reaching back to pass on the meaning of bye-gone days – are magical.   Personal twinkles transferring the essence of history between generations are rarely captured on film yet remain indelibly impressed in the mind of the beneficiary.

Vibrant memories survive in the soul as the bona fide connection between ages. Those presented here are keen to me.

I hope they appeal to you in some small way.

And may the concluding pages reveal and rekindle the spirit of self sufficiency and unassuming potential that so marked the vigor of our forefathers.

Ode To A Hay Derrick

Hail faded soldier of hay stacking past!
  Lifted to power amid creaking tension;
Your silhouette greets each passing cloud,
  In salute tossed high at load bent attention.

Be you called Wilson, A-frame or swinger,
  Or Mormon with Jackson hanging on cable,
Beaverslide, overshot or high incline slider,
  Pulleyed or bucker - your history is fable.

Hear proud relic of sun baked homestead,
  Nostalgia filled meadows thrive yet today.
Pitchforks and trip ropes are not all forgotten,
In spirit and memory, you're still making hay.

Carry on soldier. Carry on.

Remembering
The Good Olde Days

Salt Lake City

Anywhere

Cle Elum, Wash.

Denver, Colo.

Trout Creek, NV

Rupert, Idaho

Pole derricks were used extensively in Northern Nevada's Paradise Valley. A Library of Congress series called Buckaroos in Paradise highlights haying activity on the Boggio, Ferraro and Holt Ranches in the early 1900s. Numerous pictures can be viewed on line. A documentary film, narrated by Leslie Stewart, a local rancher, was shot on the 96 Ranch in 1945. It gives a real life feeling of the loose hay stacking era.

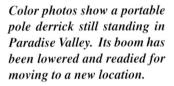

Color photos show a portable pole derrick still standing in Paradise Valley. Its boom has been lowered and readied for moving to a new location.

An Onieda County Idaho homestead near the Utah border. The braced boom derrick in the background was an early modification to the simple upright pole derrick. Heavy bracing permitted heavier loads at the expense of balance and mobility. From the Library of Congress

Crude and creaky appearances can't mask the extraordinary leverage achieved with cunning simplicity. Physics maximized with a modicum of materiel!
Quaint apparatus that really worked!

Above: Boss Wilkens' Ranch, Gilliam, County, Oregon
Below: A wheeled beaverslide from North Dakota State Historical Society files

Filmore, Utah

Twenty-six irrigation projects developed by the Bureau of Reclamation in the early 1900s converted thousands of acres of sagebrush into productive new farm land.

Cecil Barton's Jackson Fork

92

Marsh Hay and Oxen

Marsh hay, a tideland native of the east coast - particularly New Jersey - was cultivated extensively in colonial times. Harvesters and horses had to be carefully protected from mosquitos during scything and hauling. Today the plant is used primarily for seaside erosion control.

An excellent example of building a stack the hard way - purely by pitchfork. This photograph was taken on the J.E. Hinton Ranch, later to be the Imperial Ranch in central Oregon near the town of Shaniko.

Laying It Down
By Hand and By Horse

Peter Henry Emerson Series
George Eastman House Rochester, N.Y.

Pliny, The Elder, a Roman scholar, wrote a moving piece about scythes back around AD 50, some 500 years after the Persians started growing Alfasfasah. Folks later came to call the crop alfalfa but that was the only change in hay farming for a very long time. Crescenzio, another scientist from the same country - different regime - did an update in 1548 and said sickles were pretty much still in fashion and not significantly different in style than those used by his great, great, great, etc., granddad. You know, it wasn't until 1822 that Jeremiah Bailey invented a horse powered mowing machine that single-handedly laid down 10 acres in one day.

United States Library of Congress

Hay Cutters

Jeremiah was a clever chap but stayed a little too linked to the past. He welded a series of scythes to a circular frame and geared the contraption to the mower's wheels. Even hooked up a whetstone to sharpen the blades as they spun. It took a few more years before Cyrus McCormick, A.J. Purviance, Obed Hussey, Wm. Ketchum and a few other contemporaries perfected the alternating sickle to work as we now know it. They even found a better way to mechanically sharpen the sickles using a wheel mounted grinder. For some unknown reason though, nobody ever automated the loose haystack cutter needed to slice the hay after it was stacked. Perhaps an opportunity lies in waiting or does the principle of finding a need and filling it still apply here?

Gathering It In
Rakes, Slips and Loaders

Once cut, alfalfa, timothy, red clover or any of the other legumes used for feeding animals, had to be collected and brought in from the field to a common storage area. In the beginning, freshly cut stems were piled up like leaves using a simple hand rake and toted to the manger by fork or small cart. A multitude of wheeled rakes, loaders, slips, wagons and other specialized devices were invented along the way to ease the task. Like all toil, haying was fun if done in a nurturing sense. It was a kind and relaxing chore. I loved being alone with my team thinking and singing the day away in the wide open, aromatic countryside.

Uncle Matt's old Doubletree - My company logo

96

Making Hay

Hay making begins as the stem is cut. From there on the factor supreme - moisture - takes control. An old saying sums it up: *"The moisture without won't hurt you, but the moisture within will."* Another adage, possibly more familiar, advises: *"Make hay while the sun shines."*

Both relate to the essence of *'making hay'*. For hay is not what is planted and grown in the field. Hay is what is *'made'* in the curing process. It is the fodder resulting from the curing or drying of the crop after it is mown. Nature displays her favor here.

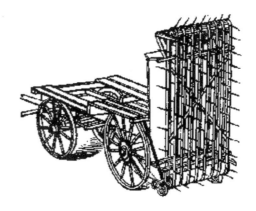

Gilliam County Oregon

Toppenish Valley Washington
One in the 'Mural-in-a-Day' series by
Kooskia, Idaho painter Robert Thomas

Sb

Utah

Stacking It Up
Imagination Meets Need

Kansas

Oregon/California

So

Tc

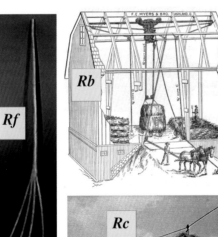
Rb

Rf

Slide stackers **either used plungers with long poles to push loads up a wooden incline or pulleys mounted near the top of the slide to pull the hay up with cables. Lifted jags dropped over the edge to the stacking surface.**
Toothed stackers **mechanically elevated loads with cradles made of long poles or teeth. Horses pulled on cable assemblies to lift the cradles.**
Boom stackers **came in a wide variety of frameworks from simple upright poles to heavily braced, wide bodied derricks fitted with long swinging booms. These were the giants of the loose hay stacking era.**
Rudimentary devices **ranged from simple labor-intensive hand forks to tripods and rope systems. They preceded more advanced systems.**

Rc

Rr

Rt

B7

Central Oregon

98

To

Wyoming

Ts

Jordan Valley, Oregon

B1

17 Ways

Folks Found To Put Up Hay

Sb	*Beaverslide*
So	*Plunger overshot*
Tc	*Combination pushrake/stacker*
To	*Back flip overshot*
Ts	*Swing-around*
B1	*Leaning mast*
B2	*Jointed or chained pole*
B3	*Wilson*
B4	*Braced boom*
B5	*Top mount pivot boom*
B6	*A-frame*
B7	*Portable boom*
Rt	*Tripod*
Rc	*Suspended cable*
Rr	*Rope pull*
Rb	*Barn trolley carrier*
Rf	*Pitch fork*

B2

Paisley, Or.

B4

Paul, Idaho

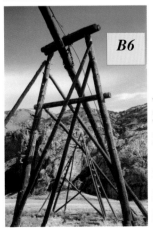

B6

ID 28 Above Mud Lake

B5

B3

99

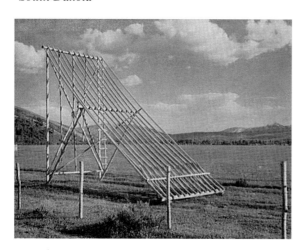

Sliders
Beaverslides and Overshots

South Dakota

Carmen Creek, Idaho

Plunger Push
Slider

Slide stackers became popular in mixed hay and grass growing locales due in part to the texture of the hay crop grown. Inclined ramps adapted to these lighter crops better than jackson forks and sling chains. Horses hitched to long poled plungers pushed cradles up wide bodied, fan-shaped styles. Open ended ramps with pulleys mounted to the upper ends of extended poles used cables to pull up the loads. Both styles were interchangeably called overshots and beaverslides though the open ended model is properly labeled a beaverslide. It got its name from a Western Montana County where it was invented in 1910. Beaverhead County is known as the land of 1,000 haystacks due to the wide proliferation of this model stacker.

North Dakota Historical Society 0032-LO-14-17

Library of Congress - Nebraska Beaverslide

6 - 12 - 6

Simple concepts enlighten.

6 - 12 - 6 is probably a riddle now but once was a common idiom. Simply stated: "If you want to eat at the table, wash up and be ready at 6:00, 12:00 and 6:00. There was work to be done and a proper time to do it.

Clear enough?

Wyoming

Beaverhead County, Montana

Toothed Lifters
Backflip Overshots and Swingers

Idaho

Nebraska

Kansas

Overshot generically described two very different stackers. One raised hay by pushing it up a ramp. It was really a slider. The type shown on this page raised hay using pulleys and cable backflipping the load over itself onto the stacking surface. To distinguish between them, I settled on calling the toothed model a back flip overshot. [To] The wide topped slider model pictured elsewhere became a plunger overshot. [So]

Paradise Valley, Nevada

Bellevue, Idaho

Buckrakes were driven rambunctiously under windrows collecting up to 1,000 pounds for delivery to waiting slides and forks of standing stackers. Steel tips protected wooden teeth though some felt pointed logs worked as well since there was no tip to break off. Bucks and combos were prone to collecting dirt, an unsavory extra.

102

Swingers

The 'Swing-Around' was perhaps the most complex of all stackers. These cleverly designed machines took on loads brought in from the field by sweep rakes and could lift and swing loads from either side. This made it possible for two sweeps to approach without interfering with each other. Cabling lifted the load and returned the cradle for another lift. An intricate spring loaded arm held the cradle arm from swinging over the stack until elevated up to stack height.

Combinations

Certainly the most bizarre of the behemoths had to be the combination push rake/stacker. These wobbly, wheeled structures could do it all - pick up a load in the field, run it to the stack yard and then crank it to elevation. Usually they parked 80-100 feet from the stack, got a load from a sweep rake and charged. A windlass hooked to the main axle did the lifting. Timing and distance was key. Below is one starting a stack in Iowa.

Baker, Oregon

Iowa

103

Pivots and A-frames
The Giants In The Field

Nampa, Idaho

Irrigon, Oregon

Vale, Oregon

Minidoka County, Idaho

*How dear to this heart are the scenes of my childhood, when fond
recollection presents them to view, the orchard, the meadow, the
deep tangled wild-wood, and every lov'd spot which my infancy
knew. The wide spreading stream, - the mill that stood near it, the
bridge and the rock where the cat-a-ract fell. The cot of my father,
the dairy house by it, and e'en the rude bucket that hung in the
well.*

**The old oaken bucket, the iron bound bucket, the moss cover'd
bucket that hung in the well.**

Samuel Woodworth 1785 - 1842

The author placed an ad in the Ruralite Magazine – a public utility newsletter catering to rural electricity users – requesting information and pictures about old time haying methods. Two excellent replies were received. I owe much to Mr. Rod Donnelly of Spray, Oregon and Mr. Ariel Bean of La Grande, Oregon for their pithy, to the point letters and kind and careful proofreading.

Background Sketches

By Rod Donnelly
Spray, Oregon

Dear David

I saw your ad in the last Ruralite expressing an interest in loose hay stacking. I don't have any articles or pictures but I do have lots of recollections, which I am willing to share with you. Perhaps you will find some thing of interest in them.

A bit of background. I was raised on a commercial sheep ranch. We ran two range bands of ewes of 1,100 head each. We tried to raise all our own hay and grain for feed. Some times we would be caught short but usually not. We dry farmed approximately 200 acres and had 23 acres irrigated from the John Day River. The dry land was seeded to various grains – rye, wheat and barley. Half of the 200 acres were seeded and the rest summer fallowed. Then, next year, the fallowed land was planted. Usually we relied mostly on rye and wheat because they could be seeded in the fall while barley was seeded in the spring and unless one had favorable weather conditions, it was more subject to failure.

We farmed with horses until the early 1940s when we started converting to tractor power. The war put a halt to that so it was in the post war years when machinery again became available that we got rid of the horses. We had 10 to 12 two horse work teams, half a dozen saddle horses and perhaps, 10 or so pack horses. These often did double duty as saddle animals. Our haying procedure was the same as every one else in the area and was as follows:

First the hay was mowed with horse drawn mowers and then was raked into long windrows with 2 horse drawn dump rakes. This latter job was usually given to a kid – Me – driving the slowest, most dependable team. Then when the hay had dried, usually a day after cutting, it was hand shocked by men with pitchforks. Depending on work progression, it could be and often was, left in these shocks for several days. If it rained and it got wet, which happened rarely, it had to be hand turned and later re-shocked after it dried out. When it came time to haul the hay in to the stacks, the shocks were hand pitched onto horse drawn hay wagons by 'field pitchers'. As a rule of thumb, one pitcher could pitch for two wagons. We usually had only two wagons in the field at a time simply because the man on the stack couldn't keep up with more than two. If the hauls were long he could stack for three, but usually, if we had

three and only one man on the stack, one of the wagons would be waiting. Also this was marginal for one man pitching in the field.

The man on the wagon had to take the hay as the pitcher threw it on and spread it out and pack it down. He couldn't simply crawl on top of it and drive away. There was quite a knack to putting a good load together so that you got on a lot of hay that wouldn't fall off on the way to the stack. Once at the stack it could be much more efficiently and quickly unloaded if it had been properly loaded to begin with.

We did almost all our unloading at the stack using a derrick pole and a Jackson fork. I have no idea how much you know about stacking equipment but there were several different kinds of stacker types – the overshot stacker, the slide stacker, the jayhawker, etc. The one we used was different from any of these. A big load of hay could be put on the stack with about 4 Jackson fork loads plus a 'clean up' fork full. Some times a load would turn up that it took more but usually, if the wagon loader did a good job, 4 or 5 were enough.

As for the stacking itself, usually one man could do it for two wagons. My Dad, and later on me, usually did the stacking. It required good strong legs and leg cramps or 'charley horses' at night were not uncommon. You start your stack by establishing a centerline and keeping the hay spread out evenly. Keep it a bit higher in the center and keeping it firmly packed – no loose spots. This is where strong legs come in. You pack it down by tromping on it. If you leave loose spots, this will leave depressions or hollows when the hay settles over time and allows water to run down. You can get rotten spots clear to the ground.

With the derrick set up we used, we built half circle stacks with the pole at the center of the circle. The boom pole angle could be changed to drop the hay closer to the center of the circle or farther out towards the rim. This was at the direction of the stacker. The hay was dumped from the fork wherever the stacker wanted it. Four or five fork loads would be pretty evenly spaced across the stack surface. The stacker had to tear apart, put in place with his pitchfork, and smooth it out as well as to tromp on it. You simply don't yell 'dump 'er' and then crawl on top of the mound and wait for more. You had to adjust the size of the stack to fit the hay supply. If I had all the hay I wanted, I liked to make stacks approximately 40 feet across and build it straight up for 10 feet or so, then bulge it slightly for another 20 feet and then take another 10 feet to top it off. When properly finished, the top would turn or shed water almost like a thatched roof. If you had left no soft spots, the hay would be bright and good after even a couple years. This might not work in the Willamette Valley but in the relatively dry climate we have here, spoilage would be minimal before it was all fed out.

Usually it was all fed the first winter but some times there would be a carry over and if so, it was fed first the following winter. A stack with a properly constructed top would have a hard, discolored crust and all the rest would be bright and good. Even the crust could be fed but the feed value was lower.

The stacks were usually fed a section at a time, cutting down through the hay with a hay knife. This left, say 3/4ths of the stack with its top still on, while you fed out another quarter. Then start on another section. If the stack was small, the whole top would probably be thrown off when you started. The hay would be pitched by hand from the stack onto a wagon and then hauled out to the feed ground. Sheep were fed twice a day, cattle once.

At least this was the way haying was done on our place and our neighbors in the pre-tractor and baler days. Modern haying methods, of course, are far more efficient and less labor intensive.

Hope the above will be of some interest to you. You asked about stacking and I have given you a lot more about other aspects of haying operation. If you have further questions that I can answer, I would be glad to do so. How come you are interested in this subject? Doing some writing?

Sincerely Rod Donnelly Spray, Oregon
June 12, 1999

Memories Of Haying In The 1930s

By Ariel Bean
LaGrande, Oregon

Early in the year before the alfalfa began to grow, my dad always had the fields spring toothed to kill weeds and cheat grass. This may have split some of the alfalfa roots but didn't seem to hurt overall production.

When the alfalfa reached about ten percent bloom, it was considered time to cut it. This was done with a horse drawn mowing machine. After some drying time, a horse-drawn barrel rake windrowed the hay. It's possible to see some of these; sometimes as lawn ornaments; sometimes parked with other unused farm machinery.

The next step was to shock the hay. This meant that men using pitchforks broke up the windrow and formed little stacks of hay about the size one could pick up and load with one attempt. How to pick up a shock of alfalfa hay properly is for the most part a lost art.

When the hay was judged to be sufficiently cured, it was ready for hauling to market or for being stacked for later use or sale. We used a slip (a flat set of boards about 8 to 10 feet wide and 12 or so long) to move hay from the field to the stacking area. The slip was loaded, usually by two men, while someone else drove the team of horses, which pulled the slip along the windrow. I had the job of driving the horses. When the loaders had put on as much hay as they figured would stay aboard, the slip was driven to the stack. Pulling the slip over the alfalfa stubs put a real shine on its bottom. The slips - we used two - had a chain from one front corner to the other to which a set of double trees were fastened. When hooking up, the harness tugs fastened to the single trees - four hookup points.

What has become known as a "Mormon" derrick was used for the stacking. What is left of a few of these can still sometimes be seen in farming areas. Ours had either logs or about four by twelve skids, which were, braced both crosswise and at an angle. In the center of the skids, a vertical post was supported

and braced. This was, at a guess, twelve to fifteen feet tall. Mounted at the top of the post was a long log boom on a swivel. The bottom end of the boom was anchored to the post by a heavy chain in such a way that the boom could move. At the tip of the boom a big pulley was mounted. Another one was mounted about where the boom and post met. A third one was at the bottom of the post and another one was at a corner of the skid, which was away from the haystack. A cable - half inch, I think - ran through these pulleys with a Jackson fork on the end of it below the tip of the boom and a single tree on the end near the skid corner.

The Jackson fork had four large tines, which held the hay as it was moved from the slip to the stack. There was a lock to hold it in the right position after it was loaded by sticking it into the hay on the slip. A trip rope ran from the lock to the man on the slip by which he could dump the fork when the stacker yelled. The rope also was the way the fork was pulled back to the hay on the slip. The derrick horse pulled the loaded fork into the air, stopped on command and then backed up to do it all over again after the fork was dumped. We always had someone walking alongside the derrick horse for control, but one of our horses got smart enough that she could almost do it by herself. The Jackson fork was not a safe piece of equipment. The tines were sharp and shiny. The stacker could be fatally stabbed, although it never happened to us.

The man on the stack had to know what he was doing to place hay on the corners, build the middle, and finally to top off the stack so that it would shed water rather than let it go down into the stack itself. Not every hay hand could do the stacking. That also is probably a lost art. When one stack was considered tall enough,

the derrick was moved over so that a new stack could be made alongside the first.

We had forty acres of alfalfa, which was about a quarter of a mile from the other big field. Six or eight horses were hitched to the skids of the derrick, the boom was tied so it couldn't swing around, and the horses pulled the derrick to the new location. To keep the wind from destroying the stacks, my dad always had a barbed wire over the top with an old post hanging from each end - and possibly a couple of cross wires and posts.

"HAY"-OUR-BIG-CROP-IN-MODOC

From the Eastman's Originals Collection, Department of Special Collections, General Library, University of California, Davis. The collection is property of the Regents of the University of California; no part may be reproduced or used without permission of the Department of Special Collections.

A lot of hay was delivered 12 miles into town by team and wagon. The hay would be loaded in the evening, and on the next morning, the man who was doing the delivery would start out by the back roads. After arriving at the delivery point, he would pitch, by hand, the hay, into the hay mow of somebody's barn, and move it all back to accommodate the load or make room for a second load. He would then drive back to the farm and reload the wagon for the next day's trip. Sunday was a day of rest for the men and teams.

One day my cousin Vic and I were shocking a couple of long windrows when we noticed that every time we moved hay into a shock, the windrow wiggled about fifteen feet further on. Finally, when we got down near the end, several skunks ran out of the end of the row. We were fortunate as they ran into a plowed field of the neighbor's rather than make a stand.

Our minimum haying crew consisted of two loaders in the field, two slip drivers, one derrick boy, one stacker, and one fork loader - a minimum of seven. Now one man with a windrower, baler and wagon can do it all in less time.

Ariel L. Bean   La Grande, Oregon   June 14, 1999

The Austrian Scythe Part Two Mowing Technique
A reprint from the Small Farmer's Journal, Summer, Vol. 19, No. 3

By Peter Vido

Perhaps less than 100 years ago there was no need to write instruction books on how to use the scythe. Just give a man a good tool and he'd be sure to make himself some hay. Oversimplified advice "just go out and swing the thing" has been given in some published sources. Early this century it may have sufficed. Today, I feel it does not, for there is a more modern, and distorted version of the swinging that has already become somewhat ingrained.

By far the majority of casual scythe users I've seen in this country swing the scythe as if it was a machete on a long handle - hacking forcefully at the grass. A scythe, be it the American or Austrian style, is primarily a slicing, rather than hacking, tool. The Americans of old well knew this and used it as such. Some are still around and if you can get one to do a live demonstration for you, by all means do so.

My notion of how the degeneration may have happened is that many of the old scythes left hanging on a nail were dull and rusty, their edge worn away to have now altogether too thick a bevel to sharpen easily. Some stroking with a hand stone just wouldn't do it and the old large grindstones, and the men who could use them were getting rare. A dull scythe won't slice grass, but it will, however inefficiently, hack some of it off if used with enough force. To gather sufficient momentum the heavy tool was backed away (and lifted up in the process) to, take a run at its target. Grass fell away, scattered in an un-orderly heap, but it was better than not cutting it at all. This technique continues because a scythe is no longer a tool of much importance. Even a fool would not try to cut, say an acre of hay in this way, or if he did he'd soon change his mind about romantic old ways of doing things.

This hacking style is not only much slower, it cannot leave a lawn or a hayfield nice and even. It is also difficult, if not impossible, to be very accurate using it around small trees and bushes. One "advantage" in using the American pattern of scythe is that one does not have to bend as low as the pioneers of old - and still lay some grass down. Perhaps we're not supple enough to bend over much nowadays or do not know

any better. I hope it is the latter, for it's more easily corrected.

Here is a little story to illustrate just how ingrained this hacking technique can be. One of our friends, who last winter read *The Scythe Book,* bought one of the Austrian scythes. He'd been playing around with it for a few days when I came to visit him. In spite of the very lucid explanation in the book, he wasn't altogether satisfied with how the grass responded. His blade was not razor sharp, his snath too short and the nibs too loose. Nibs are the stubby handles mounted on the snath. The lowest one should be positioned about the length of the blade from the bottom of the snath. The upper nib is placed the length of your own forearm above that. For most people, that means you'll be bent over a bit when you swing the tool, but power for scything comes mostly from the twisting of the torso, not from the arms.

He did not quite realize the importance of the adjustments, but partially, the trouble lay in his technique of still lifting the blade before a stroke and doing a sort of hybrid of the two types of swings. While I was doing a demonstration for him, another friend, a seasoned homesteader, also with experience in using the American scythe, came by and watched.

After awhile he wanted to try this shiny new tool and, in spite of all the watching, did just what I was trying to have my friend unlearn ... now the man who drove me to see my friend was there as well with his new, city-born wife who may have - the day before - not even known what that contraption we were playing with was for. In her "sweet 16" - like voice she suddenly called, "Let me try it, too!" Being polite, my friend handed it over. I thought to myself as she took the scythe, "We'd better all stand far enough away lest she cut our throats with it..."But, by gosh, she did a few perfect semicircles, just like an old pro and put both of them seasoned homesteaders to shame.

The moral of the story is that it may be better to get a green horse than a spoiled one with some undesirable ingrained habits, even if the latter one may do some pulling for you right away. Or, if the hacking style is one you've been using, try to forget everything you knew about using a scythe before you approach the grass with your new Austrian blade.

After having helped various people to do the swinging, I see that there are three basic, almost universal, habits to unlearn. The first is the lifting of the blade up and back into the air (often a foot or more) before the blade enters the grass. The second is taking too much of a bite in the forward depth of the grass and thirdly, not finishing the cut but instead lifting the point of the blade too soon. I can imagine a "blade tethering device" which would hold it by a cord, maybe 2" long, to the ground and yet allow it to move only in a semicircle around the mower. Try practicing for awhile with closed eyes as if your blade is tethered to the ground, twisting at your waist simultaneously as you "swing the thing" until your hands and body get imprinted with the motion. Bodies sometimes learn faster than minds. There is no need to lift the blade at the beginning of a stroke – a sharp blade will cut from a standing start and lifting it only distorts the horizontal accuracy you are trying to poise yourself for.

Now these blades are only meant to cut 4" to 6" deep at a stroke. While more is possible with light grass stands, clean grain fields and experienced hands, it would be better to start with the lower limit, or even at first just a 2" slice at a time. You can this way watch the blade better, keeping it parallel along the whole stroke. As you are finishing try to exaggerate the inward motion with the point to make sure all the grass is cut. There is a tendency to worry about not digging the tip into the ground, hence the elevation of the blade as it nears the end of the cut. You will do a little ground digging in the learning process unless you are one of the "all naturals". But not to worry; a good quality blade will take some abuse if the force behind it is not too great. The curvature of the Austrian blade is such as to discourage it and I find most folks need to be reminded to keep the point low so as not to leave high ridges along the end of the cut.

As for the strength applied, it is much less than you'd expect, if your blade is very sharp and technique correct. So instead of energetic swinging, apply some of the energy in extra sharpening and you'll have enough left over for going fishing or dancing after the grass is cut.

Peter Vido   New Brunswick, Canada

Mr. Vido has traveled and written extensively on scything. He also conducts seminars and workshops on scything in America and in Europe, particularly his homeland Austria.

Upon reading this Mr. Bean of La Grande, Oregon commented that he'd *even consider dancing* before ever picking up a scythe again, a tool he has left hanging for years.

Hay Stackers by State, Type and County

State	Stacker Type	County and City	Hwy/MP/Address	Ranch Name
California	B2 Suspended Poles	Lassen, Chilcoot	US 395	
	Sb Beaverslide	Modoc, Eagleville	CA 447 19 S	
	Ts Swing-around	Modoc, Eagleville	CA 447 20 S	
Canada	B1 Leaning Mast	British Columbia, Kelowna		
	B4 Braced Boom	British Columbia, Kelowna		
	Tc Combination	British Columbia, Kelowna		
Colorado	B6 Single A-Frame	Montrose, Montrose	US 550	
	Sb Beaverslide	Jackson, Walden	CO 14	Joe Green/J. Lawrence
	So Plunger Overshot	Jackson, North Park	CO 125	Humphries and Gray
	Ts Swing-around	Ouray, Ridgway	US 550	
Idaho	B2 Suspended Poles	Onieda, Malad City	US 91	
	B3 Wilson Derrick	Franklin, Preston	US 91	
		Minidoka, Paul	I 84 453 S 850 W	Carl Schrock
	B4 Braced Boom	Canyon, Nampa	ID 55 8 W	
	B5 Full Braced Pivot	Ada, Meridian	ID 55	
		Ada, Meridian	I 84 41	
		Ada, Parma	I 84 18	
		Bannock, Pocatello	US 91	
		Bonneville, Idaho Falls	US 91	
		Canyon, Caldwell	I 84 29	
		Canyon, Caldwell	US 30	
		Cassia, Burley	I 84 209	
		Cassia, Burley	I 84 300 S 120 E	
		Cassia, Burley	I 84 198	
		Cassia, Burley	I 84 600 S 100 W	
		Custer, Mackay	US 93 129	
		Elmore, Mountain Home	US 20 6 N	
		Franklin, Preston	US 91	
		Jerome, Hansen	I 84 180	
		Lincoln, Dietrich	ID 24	
		Madison, Rexburg	ID 191	
		Minidoka, Paul	I 84 105 S 360 W	A. Hanson
		Minidoka, Paul	I 84 90 N 760 W	Shanner
		Minidoka, Paul	I 84 200 N 546 W	Aldo Dallolio
		Minidoka, Paul	I 84 200 N 650 W	
		Minidoka, Paul	I 84 150 S 950 W	
		Minidoka, Rupert	I 84 375 W Baseline	Joe Bott

Hay Stackers by State, Type and County

State	Stacker Type	County and City	Hwy/MP/Address	Ranch Name
Idaho	B5 Full Braced Pivot	Minidoka, Rupert	I 84 150 N 200 W	Ray Bott
		Minidoka, Rupert	I 84 105 S 307 W	Moller family
		Minidoka, Rupert	I 84 200 N 350 W	Herman Bott
		Minidoka, Rupert	I 84 275 N 160 W	Frank L. Orr
		Minidoka, Rupert	I 84 115 S 280 W	Moller family
		Minidoka, Rupert	I 84 195 N 400 W	Rudy Bonadiman
		Minidoka, Rupert	I 84 140 N 400 W	Pete Wakewood
		Minidoka, Rupert	I 84 100 N 213 W	Louis Bott
		Minidoka, Rupert	I 84 430 S 100 W	Smith Family
		Minidoka, Rupert	I 84 145 S 307 W	Moller family
		Minidoka, Rupert	I 84 200 N 325 W	Steven Walters
		Minidoka, Rupert	I 84 95 S 370 W	Moller family
		Minidoka, Rupert	I 84 60 S 380 W	Moller family
		Onieda, Malad City	US 91	
		Owyhee, Grandview	ID 78 8 W	
		Owyhee, Grandview	ID 78 9 W	
		Owyhee, Grandview	ID 78 10 W	
		Owyhee, Grandview	ID 78 7 W	
		Owyhee, Riddle	ID 51 1 N	
		Payette, Fruitland	I 84 3	
	B5 Half Braced Pivot	Canyon, Murphy	ID 78 1 E	
		Canyon, Nampa	ID 45 2 S	
		Canyon, Nampa	ID 55 2 W	
		Elmore, Mountain Home	ID 78 20 S	
		Payette, Payette	ID 52 7 E	
	B6 Single A-Frame	Bannock, Pocatello	US 91	
		Bingham, Blackfoot	I 15 MP 99 3 N	
		Blaine, Carey	US 93	
		Bonneville, Idaho Falls	US 91	
		Butte, Arco	US 93 1 S	
		Cassia, Burley	I 84 500 S 450 W	
		Cassia, Declo	I 84 217	
		Cassia, Rupert	I 84 225 N 765 E	Rino Grisenti
		Custer, Clayton	ID 75 3 W	
		Elmore, Bliss	I 84 140	
		Franklin, Preston	US 91	
		Fremont, St. Anthony	ID 191	
		Gem, Emmett	ID 52 2 N	
		Gooding, Bliss	I 84 MP 139	

Hay Stackers by State, Type and County

State	Stacker Type	County and City	Hwy/MP/Address	Ranch Name
Idaho	B6 Single A-Frame	Gooding, Wendell	I 84 156	
		Gooding, Wendell	I 84 162	
		Jefferson, Mud Lake	ID 33 46 E	
		Jefferson, Mud Lake	ID 33 48 E	Wild Life Refuge
		Lemhi, Lone Pine	ID 28 45	Waggoner Ranch
		Lemhi, North Fork	US 93 3 S	
		Madison, Rexburg	ID 191	
		Minidoka, Paul	I 84 340 S 850 W	Richard Graff
		Minidoka, Paul	I 84 95 N 750 W	Domingo Eguilior
		Minidoka, Rupert	I 84 380 S Meridian	Forest Douglas
		Minidoka, Rupert	I 84 226 S 100 E	Barbara Studer
		Minidoka, Rupert	I 84 200 N 100 W	Robert Dockter
		Minidoka, Rupert	I 84 200 N 376 W	Albert Asson
		Minidoka, Rupert	I 84 300 N 148 W	Gary Schorzman
		Minidoka, Rupert	I 84 100 N 225 W	John R. 'Bob' West
		Minidoka, Rupert	I 84 300 S 200 E	
		Minidoka, Rupert	I 84 100 N 230 W	Cecil Barton
		Minidoka, Rupert	I 84 100 E Baseline	Lynn Manning
		Minidoka, Rupert	I 84 301 N 200 W	Jerry Grubaugh
		Minidoka, Rupert	I 84 201 N 400 W	Richard Schenk
		Minidoka, Rupert	I 84 200 N 260 W	Asson Homestead
		Onieda, Malad City	US 91	
		Twin Falls, Buhl	US 30 2 E	
		Twin Falls, Hagerman	US 30 5 E	
	B6 Double A-Frame	Ada, Star	ID 44 2 W	
		Bear Lake, Franklin	US 91	
		Bonneville, Swan Valley	ID 26 5 W	
		Lemhi, Lone Pine	ID 28 45	Waggoner Ranch
		Minidoka, Paul	I 84 200 S 879 W	Steve Roberts
	Sb Beaverslide	Cassia, Declo	I 84 3 E	Dewey Family
		Lemhi, Gibbonsville	US 93 5 N	
		Lemhi, Leadore	ID 28 95 E	
		Lemhi, Leadore	ID 28 94 E	
		Lemhi, Leadore	ID 28 80 E	Oxbow Ranch
		Lemhi, Salmon	ID 28 5 E	
		Lemhi, Salmon	US 93 5 N	Carmen Creek
	To Backflip Overshot	Bear Lake, Montpelier	US 30 437	
		Blaine, Bellevue	ID 75 10 S	J. K. Molyneux

Hay Stackers by State, Type and County

State	Stacker Type	County and City	Hwy/MP/Address	Ranch Name
Idaho	Ts Swing-around	Cassia, Rogerson	US 93 22	
		Custer, Challis	US 93 253	
		Cassia, Elba	ID 77 450 E 435 S	Delbert Glaesemann
		Lemhi, Leadore	ID 28 44 E	Waggoner Ranch
		Lemhi, Salmon	ID 28 4 E	Snook Ranch
		Minidoka, Rupert	I 84 325 S 225 E	
		Power, American Falls	ID 39 5 N	Haderlie's River Ranch
Iowa	Tc Combination	Jasper,		Kimberly Farm
Kansas	Tc Combination	Scott,		
Montana	B2 Suspended Poles	Wheatland, Harlowton	US 12	Winnecoop Ranch
	B5 Full Braced Pivot	Beaverhead, Dillon	I 15 77	
	Sb Beaverslide	Beaverhead, Big Hole	MT 278	David Hirschy
		Beaverhead, Dillon	I 15 60	
		Beaverhead, Grant	MT 324 5 W	Bar Double T
		Beaverhead, Grant	MT 324 7	
		Beaverhead, Grant	MT 324 6	
		Beaverhead, Grant	MT 324 15 W	
		Beaverhead, Grant	MT 324 5	
		Beaverhead, Jackson	MT 278	
		Beaverhead, Wisdom	MT 278	
		Beaverhead, Wisdom	MT 278	
		Beaverhead, Wisdom	MT 278	
		Granite, Hall	MT 1 54	
		Granite, Philipsburg	MT 1 35	
		Granite, Philipsburg	MT 1 36	
		Jefferson, Elliston	US 12 18	
		Powell, Avon	US 12 5 W	
		Powell, Avon	MT 141 15 N	Charles & Rita Gravely
		Powell, Avon	US 12 12	John Beck
		Powell, Avon	US 12	Joyce Bignell
		Ravalli, Hamilton	US 93 10 S	
		Silver Bow, Wise	MT 43 57	
	So Plunger Overshot	Pondera, Dupuyer	US 89	Beaverhead Trading
		Powell, Elliston	US 12 18	
		Teton, Choteau	US 89 8 W	
	Tb Buckrake	, Hysham		Ed McCormick
	To Backflip Overshot	Judith Basin, Stanford	US 87 2 W	
	Ts Swing-around	Granite, Drummond	MT 1 61	

Hay Stackers by State, Type and County

State	Stacker Type	County and City	Hwy/MP/Address	Ranch Name
Nebraska	To Backflip Overshot	Buffalo, Kearney	NE 40	Watson Ranch
		Buffalo, Kearney	NE 40	Empire Ranch
		Lancaster, Lincoln	MT 79	
		Lincoln, North Platte	I 80 180	
Nevada	B2 Suspended Poles	Clark, Bunkerville	NV 170	
		Eureka, Eureka	US 50 11 W	Roy Risi Ranch
		Eureka, Frontier	US 50	
		Humboldt, Paradise Valley	NV 290 2 N	Joe Sicking
		Humboldt, Paradise Valley	NV 290	Paradise
		Humboldt, Winnemucca	US 95	Boggio Family
	B3 Wilson Derrick	Humboldt, Paradise Valley	NV 290 2 S	Tom & Dave Cassinelli
	B4 Braced Boom	Humboldt, Paradise Valley	NV 290	
		Humboldt, Paradise Valley	NV 290	Pasquale Family
		Humboldt, Paradise Valley	NV 290	Holt Ranch
	B6 Single A-Frame	Humboldt, Winnemucca	I 80 176	
	B6 Double A-Frame	Clark, Bunkerville	NV 170	
	Sb Beaverslide	Elko, Contact	US 93 45	
		Elko, Deeth	I 80 328	
		Elko, Deeth	I 80 333	
		Elko, Halleck	I 80 328	
		Pershing, Lovelock	I 80 105	
	Ts Swing-around	Elko, Elko	NV 225 15 N	
		Elko, Owyhee	NV 225 1 N	
		Elko, Tuscarora Junction	NV 225 15 N	
		Elko, Tuscarora Junction	NV 225 20 N	PX Cattle Co
		Eureka, Eureka	NV 278 62 N	
		Eureka, Eureka	NV 278 73 N	
		Eureka, Frontier	US 50	
		Humboldt, Paradise Valley	NV 290	Kenny Buckingham
		Lander, Austin	NV 305 30 N	
No. Dakota	Sb Beaverslide	,		
	Tc Combination	,		Y. O. Ranch
	To Backflip Overshot	,		
Oregon	B2 Suspended Poles	Gilliam, Condon	OR 206	Boss Wilkins
		Lake, New Pine Creek	US 395	
		Wheeler, Spray	OR 207	Rod Donnelly
	B3 Wilson Derrick	Malheur, Jordan Valley	US 95 5 W	

Hay Stackers by State, Type and County

State	Stacker Type	County and City	Hwy/MP/Address	Ranch Name
Oregon	B4 Braced Boom	Lake, Paisley	US 395 5 S	
		Lake, Paisley	US 395 3 S	
		Lake, Paisley	US 395 4 S	
		Lake, Paisley	US 395 10 N	
		Umatilla, Hermiston	US 395	
	B5 Full Braced Pivot	Deschutes, Sisters	US 20 5 E	William Meek
		Morrow, Irrigon	US 730	
		Union, LaGrande	OR 237	Ariel Bean
	B6 Single A-Frame	Harney, Fields	OR 205 10 N	
		Malheur, Huntington	I 84 350	
	B7 Portable A	Wheeler, Mitchell	US 26	C. H. Burgess
	Sb Beaverslide	Grant, Seneca	US 395	Oliver Ranch
	So Plunger Overshot	Lake, New Pine Creek	US 395	
	Ts Swing-around	Baker, Baker	US 30 5 N	
		Baker, Baker	OR 7 3 S	
		Baker, Baker	I 84 3 E	
		Baker, Durkee	I 84 329	Jean Bunch
		Baker, New Bridge	OR 86 34 E	
		Deschutes, Redmond	Or 126 3 W	
		Grant, John Day	US 26 168	
		Grant, Seneca	US 395 1 N	
		Jefferson, Sisters	Trail 20 N	
		Lake, New Pine Creek	US 395	
		Malheur, Ironside	US 26 230	
		Malheur, Jordan Valley	US 95 5 N	
		Malheur, Jordan Valley	US 95 10 N	
		Union, Elgin	OR 82 10 E	
So. Dakota	Rf Pitch Fork	Hyde, Highmore		Richard Harter
	Rr Rope Pull	,		Humphries and Gray
		Hyde, Highmore		Richard Harter
	Sb Beaverslide	Jones, Murdo	I 90 170	1880 Town
	Tb Buckrake	Hyde, Highmore		Richard Harter
	To Backflip Overshot	, Sioux Falls		Farrar Photo Collection
		, Wall	I 90 110	
Utah	B2 Suspended Poles	Beaver, Beaver	US 91	
		Iron, Parowan	US 91	
		Iron, Parowan	US 91	
		Millard, Fillmore	US 91	

Hay Stackers by State, Type and County

State	Stacker Type	County and City	Hwy/MP/Address	Ranch Name
Utah	B2 Suspended Poles	Millard, Kanosh	US 91	
		Millard, Scipio	US 91	
		Salt Lake, Salt Lake City	US 91	
		Salt Lake, Salt Lake City	US 91	
		Utah, Provo	US 91	
		Utah, Provo	US 91	
		Washington, St. George	US 91	
	B3 Wilson Derrick	Box Elder, Brigham City	US 91	Dean l. May
		Iron, Cedar City	US 91	
		Juab, Nephi	US 91	
		Utah, Utah Lake	US 91	
	B4 Braced Boom	Beaver, Beaver	US 91	
		Iron, Parowan	US 91	
		Juab, Nephi	US 91	
		Millard, Cove Fort	US 91	
		Millard, Fillmore	US 91	
		Millard, Scipio	US 91	
		Sevier, Joseph	US 89	
		Utah, Provo	US 91	
	B5 Full Braced Pivot	Beaver, Beaver	US 91	
		Box Elder, Brigham City	US 91	
		Cache, Cache Valley	US 89	
		Daggett, Manila	UT 150	
		Iron, Parowan	US 91	
		Juab, Nephi	US 91	
		Millard, Cove Fort	US 91	
		Millard, Fillmore	US 91	
		Salt Lake, Salt Lake City	US 91	
		Utah, Provo	US 91	
	B6 Single A-Frame	Box Elder, Snowville	US 30 5 W	Rose Ranch
		Salt Lake, Salt Lake City	US 91	
		Utah, Utah Lake	US 91	
	Sb Beaverslide	Daggett, Manila	UT 44	
		Uintah, Vernal	UT 44	
		Uintah, Vernal	UT 40	
	Ts Swing-around	Carbon, Heber	US 40 1 N	
Washington	B5 Full Braced Pivot	Columbia, Starbuck	US 12 1 W	Accuntius Ranch
		Klickitat, Goldendale	US 97	

119

Hay Stackers by State, Type and County

State	Stacker Type	County and City	Hwy/MP/Address	Ranch Name
Washington	Rc Suspended Cable	Columbia, Starbuck	US 12 1 W	Accuntius Ranch
	Tc Combination	Columbia, Starbuck	US 12 1 W	Accuntius Ranch
	Ts Swing-around	Walla Walla, Gardena	US 12 3 W	Byrnes Ranch
Wyoming	B4 Braced Boom	Bighorn, Greybull	US 20 1 W	
		Bighorn, Greybull	US 20 1 W	Dirty Annies
	Sb Beaverslide	Carbon, Rawlins	I 80 240	
		Teton, Jackson Hole	US 191	
	So Plunger Overshot	Albany, Laramie	I 80 298	
	Tb Buckrake	Sweet Water, Salt Wells	US 30	DeBortoli Family
	To Backflip Overshot	Bighorn, Frannie	US 310 1 N	
		Bighorn, Otto	WY 30 1 W	
		Fremont, Dubois	US 26	
		Lincoln, Freedom	US 89 1 W	

1. Stacker names were expanded slightly to help in distinguishing special attributes yet generally retaining labels common to the time.   Most coding is of the author.

2. All sightings of the author were made from the highway designated with the exception of those observed on back country farm roads where a rural addressing system was in use. In those cases, the nearest major highway is also listed. Mileposts are shown where available. Numbers with direction letters (34 E, 5N) indicate the approximate distance the stacker was observed from the indicated city.

3. Ranch names were observed on sign posts or taken from data supplied by historical archives or reference books from which the sighting was obtained.

Note: Details regarding every sighting are included in a common data base and can be reported in numerous formats. This report is sorted first by state, next by stacker class and further by county and city. All sightings found are included in this report. It is possible to limit the entries by area (state or county) or by one, two or several kinds of stackers permitting a wide variety of useful reports.

Words Peculiar To Haying

cock (*kòk*) *noun* A cone-shaped pile of straw or hay. **cocked, cock·ing** *verb, transitive* To arrange straw or hay into piles shaped like cones.

bind·er (*bìn der*) *noun* A machine that reaps and ties grain. An attachment on a reaping machine that ties grain in bundles.

bunch (*bùnch*) *noun* A group of things growing close together; a cluster or clump: *grass growing in bunches. Verb* **bunched, bunch·ing, bunch·es** *verb, transitive* To gather together into a group.

cor·ru·gate (*kôr e-gât´*) *verb* **cor·ru·gat·ed, cor·ru·gat·ing, cor·ru·gates** *verb, transitive* To shape into folds or parallel and alternating ridges and grooves. Tiny parallel ditches, also called rills, scratched into the soil to direct the flow of irrigation water across a field.

dou·ble·tree (*dùb´el-trê*) *noun* A crossbar on a wagon or carriage to which two whiffletrees are attached for harnessing two animals abreast.

fod·der (*fòd´er*) *noun* Feed for livestock, especially coarsely chopped hay or straw.

for·age (*fôrîj, fòr*) *noun* Food for domestic animals; fodder. The act of looking or searching for food or provisions.

jag (*jàg*) *noun* A small load or portion.

leg·ume (*lèg´yōōm´*) *noun* A pod, such as that of a pea or bean, that splits into two valves with the seeds attached to one edge of the valves. A plant of the pea family.

lo·ess (*lo´es, lès, lùs*) *noun* A buff to gray windblown deposit of fine-grained silt or clay.

mow (*mou*) *noun* The place in a barn where hay, grain, or other feed is stored. A stack of hay or other feed stored in a barn.

mow (*mo*) *verb* To cut down grass or grain with a scythe or a mechanical device.

pawl (*pôl*) *noun* A hinged or pivoted device adapted to fit into a notch of a ratchet wheel to impart forward motion or prevent backward motion.

reap·er (*rê per*) *noun* One that reaps, especially a machine for harvesting grain crops.

scythe (*sìth*) *noun* An implement consisting of a long, curved single-edged blade with a long, bent handle, used for mowing or reaping.

sick·le (*sîk´el*) *noun* An implement having a semicircular blade attached to a short handle, used for cutting grain or tall grass. The cutting mechanism of a reaper or mower.

shock (*shòk*) *noun* A number of sheaves of grain or hay bunched up in a field for drying. **shocked,** *verb, transitive* To gather into shocks.

sling (*slĭng*) *noun*   A looped rope, strap, or chain for supporting, cradling, or hoisting something.

stan·chion (*stăn′chen, -shen*) *noun*   A framework consisting usually of two vertical bars, used to secure cattle in a stall.

swath (*swŏth, swôth*) also **swathe** (*swŏth, swôth, swâth*) *noun*   The width of a scythe stroke or a mowing-machine blade.   A path of this width made in mowing. The mown grass or grain lying on such a path.

ted (*tĕd*) *verb, transitive*   **ted·ded, ted·ding**. To strew or spread newly mown grass for drying. ***Regional Note:*** In 15th-century England the verb *ted* meant to spread newly cut hay to facilitate its drying. In the mid-19th century an American inventor produced a machine to ted the hay automatically and called it a *tedder.* Since modern English is inclined to make verbs out of nouns meaning implements or machines, the noun *tedder* became a verb with the same meaning as the original word *ted.*

thill (*thĭl*) *noun*   Either of the two long shafts between which an animal is fastened when pulling a wagon.

tromp (*trŏmp*) *verb* **tromped, tromp·ing, tromps** *Informal. verb, intransitive* To walk heavily and noisily; tramp. To apply heavy foot pressure on something.

whif·fle·tree (*hwĭf′el-trê, wĭf*) *noun*   The pivoted horizontal crossbar to which the harness traces of a draft animal are attached and which is in turn attached to a vehicle or an implement. Also called *singletree, swingletree and whippletree.* ***Regional Note:*** *Whiffletree,* a term primarily used in the northeast United States, is derived from an older term *whippletree,* which is used in the Upper Northern states farther to the west. The fact that *whiffletree,* the newer term, is used in the Northeast, the older dialect area, illustrates the process of linguistic change. Even as the older word *whippletree* was spreading westward into a new dialect area, it was evolving into something different— *whiffletree*— in the area where it originated, as if the older dialect area were somehow trying to keep a step ahead.

wind·lass (*wĭnd′les*) *noun*   Any of numerous hauling or lifting machines consisting essentially of a horizontal cylinder turned by a crank or a motor so that a line attached to the load is wound around the cylinder.

wind·row (*wĭnd′rō′*) *noun*   A row, as of leaves or snow, heaped up by the wind. A long row of cut hay or grain left to dry in a field before being bundled.

wind·rowed *verb, transitive* To shape or arrange into a windrow.

> **Glossary entries were taken from *The American Heritage Dictionary of The English language, Third Edition.***

Books And Periodicals

I wish to applaud and credit the authors and editors of the following books and periodicals for contributions made to this one. Their efforts were important and enjoyable resources.

Aultman, Miller & Co., Akron, Ohio; The *New Buckeye Mower.* A circa 1881 advertising pamphlet.

Austin E. Fife and James M. Fife; *Hay Derricks of the Great Basin and Upper Snake River Valley.* University of Utah Press 1985. www.upress.utah.edu.

R. A. Oakley and H.L. Westover; *U.S. Department of Agriculture, Farmers Bulletin No. 1283, How To Grow Alfalfa.* Issued December 1922; Revised January 1928.

Circular prepared by United States Reclamation Service. *Irrigation Projects of the U.S. Reclamation Service.* Issued by Government Printing Office in 1916.

F. B. Morrison; *Feeds and Feeding, A Handbook For The Student And Stockman.* The Morrison Publishing Company, Hammond Press 1936.

Burt C. Buffam, M.S. and David Clement Deaver; *Sixty Lessons in Agriculture.* American Book Company; Copyright, 1913 in Great Britain.

F. E. Myers & Bro.; *Catalogue No. 54 with price list of Pumps and Hay Tools.* Published circa 1906, Ashland, Ohio. USA. www.pentairpump.com

W. R. Humphries and R. B. Gray; *Partial History of Haying Equipment.* U.S. Department of Agriculture, Information Series No. 74, 1949.

L. A. Reynoldson; *Hay Stackers and Their Use.* U.S. Department of Agriculture, Farmers' Bulletin No. 1615, November 1929.

Clyde Walker and Arnold Egbert; *A Boom-Type Stacker.* Oregon State University Archives Extension Circular 480, Feb. 1946.

Peter Vido; *The Austrian Scythe, Part Two, Mowing Technique.* Small Farmers Journal, Vol 19, No 3.

A Pictorial History of Gilliam County Oregon Copyright 1983 Gilliam County Historical Society

Snapshots in Time; A *Pictorial History of Early Mini-Cassia.* South Idaho Press; 1998.

Historical Societies And Web Sites

The Internet is an incredible aid to research. My heartfelt thanks to the contributors and producers of public and private sites and for gracious permission to reprint key items.

Library of Congress; Washington DC
American Folklife Center www.loc.gov/
American Memory Historical Collections

National Agricultural Library, USDA; Beltsville, Maryland. www.nal.usda.gov

Economic Research Service, USDA; Washington, DC. www.ers.usda.gov

Natural Resources Cons. Service, USDA; Washington, DC. http://plants.usda.gov/

National Archives and Records Admin. Washington, DC. www.nara.gov/

General Library, Univ. of California, Davis; Davis, CA. www.lib.ucdavis.edu/

University of Idaho Library; Moscow, ID. www.uidaho.edu/

Oregon State University; Corvallis, OR. www.orst.edu

Western State Historical Societies

Historical societies provided much valuable information and encouragement. My appreciation goes especially to the following:

Washington State Historical Society
1911 Pacific Avenue, Tacoma, Wash. 98402
253-272-3500

Oregon Historical Society
1200 SW Park Avenue, Portland, Oregon
97205   503-222-1741

California Historical Society
678 Mission Street, San Francisco, Calif.
94105   415-357-1848
www.californiahistoricalsociety.org/

Idaho State Historical Society
1109 Main Street, Ste 250, Boise, ID 83702
208-334-2682 www2.state.id.us/ishs

Nevada State Historical Society
1650 North Virginia St, Reno, NV 89503
775-688-1190

Utah State Historical Society
300 South Rio Grande, Salt Lake City, UT.
84101   801-533-3500

Montana State Historical Society
225 North Roberts, Helena, Montana 59620
406-444-2694

Wyoming State Historical Society
1740H Dell Range Blvd, Cheyenne, WY 82009

Colorado Historical Society
1300 Broadway, Denver, Colorado 80203
303-866-3682

North Dakota State Historical Society
612 East Boulevard Avenue, Bismarck, North
Dakota 58505-0830 www.state.nd.us/hist/

South Dakota State Historical Society
900 Governors Drive, Pierre, South Dakota
57501-2217   605-773-3458

Kansas State Historical Society
6425 SW Sixth Avenue, Topeka, Kansas 66615
785-272-8681 www.kshs.org

Nebraska State Historical Society
1500 R Street, Lincoln, Nebraska 68501
www.nebraskahistory.org/

Author Photos And Comments

To avoid duplication, not all photos are cited as to source within the book text. The reader will be able to identify similar presentations and can properly assume credit is due the source referenced in a nearby citation.

Color photographs not otherwise identified as to source were taken by the author. The poems, *Remembering* and *Ode To A Hay Derrick* were also written by the author.